THE

HYDROGEN CONNECTION

ENERGY INDEPENDENCE, BLOCKCHAIN TECHNOLOGY, PEACE, AND PROSPERITY

DON BONGAARDS P.E.

The Hydrogen Connection
Copyright © 2022 by Don Bongaards

Published by

Illumify Media Global
www.IllumifyMedia.com
"Let's bring your book to life!"

Library of Congress Control Number: 2022909408

Paperback ISBN: 978-1-955043-77-9

Cover design by Deborah Lewis
Printed in the United States of America

I dedicate this book to my daughter, Nena Harvath, and my grandsons Britt, Stewart, Johnathan, and Evan. May they live long, healthy, prosperous, and happy lives.

CONTENTS

INTRODUCTION

L et me begin with a word of caution. While elimi-
nating fossil-fuel-derived energy may at first sound
like a good idea, it's probably a bad idea if done too
quickly. Near term worldwide geopolitical interactions
could very well upset the apple cart in terms of economic disasters and possibly war. Therefore, keep in
mind, as you read this book, that what I'm about to
propose is a long-term, sixty-plus-year purposeful and
gradual plan.

As you read this book, you should know that I'm
an eighty-year-old retired engineer who has been in-
spired by this quote by George Bernard Shaw and later
made famous by John F. Kennedy: **"Some people see
things the way they are and say why. I dream of
things that never were and say, 'why not?' "**

The Disney Corporation invented a word that sums up this quote: *imagineering*. **"It's thinking outside the box"** and by thinking outside the box some amazing things might result. That's what this book is about. It's my imagineered future that might result as we transition from fossil fuels to hydrogen. This book is not meant to be technical. Instead, it's about ideas that almost anyone can understand.

The hydrogen revolution is well underway and, as you will see, it could hopefully act as a catalyst for changing our world in a positive way. Hydrogen can be made from our almost unlimited amount of water and sunshine. In this regard, almost anyone can make hydrogen without being subject to a centralized fossil fuel industry. Consider the growing number of poor and destitute people in the world. What if they could profit from making and selling hydrogen? How about these poor and destitute people participating in the benefits associated with the modern industrialized world? How about the result of these happenings being world peace and prosperity? Although this may sound outlandish and farfetched, just remember as you read this book, *"I dream of things that never were and say, 'why not?' "*

In 1965 and 66 I served with an Army civil affairs team in Vietnam. When I saw the extreme poverty of people who were living in villages and hamlets I was appalled. When you see it firsthand you quickly become thankful for growing up in the good old US of A. I don't think many Americans really understand how destitute most of the world is. Even when we see it in movies or television broadcasts, we tend to ignore unpleasant facts so we can avoid becoming depressed. My compassion for these destitute people has caused me to think about solutions that might help. However, the solutions I propose in this book require a dedicated effort; lots of money; and will impact non-renewable resources, food, and water.

My solutions begin with the premise that not everyone can make and sell gasoline, but almost everyone has access to water, sun, and some land to make and sell hydrogen. That is **"The Hydrogen Connection."** It's a connection between a world that's in the process of transitioning from fossil fuels to hydrogen fuel, which could be dramatically changed for the better. It's a world in which every country would be energy independent. It's also a world in which no country would be held hostage to non-renewable resources and/or the food and water of another country. It's a world in which one country would not take advantage of cheap labor in another country because cheap labor will no longer exist. It's a world in which there is a deterrent for one country to go to war with another country. While all of what I've just said may sound like pie-in-the-sky thinking, let's begin a journey into the future to understand why the worldwide hydrogen revolution is taking place now.

But wait a minute, you might ask. What is this blockchain technology, and how can it play a part in world peace and prosperity? Simply put, blockchain technology is a decentralized internet data control system. It's like an interplanetary filing system that can't be erased or destroyed. In other words, all input data is individually, not centrally, controlled. Put another way, your bank accounts, investments, property records, health care records, wills, and any other personal information in the blockchain is under your personal control and cannot be changed without your permission. While blockchain technology has the potential to provide world peace, it can't do it by itself. *Oh great, now you tell me after I've bought this book!!* you might be thinking. Well, calm down. Notice I said "by itself." What I'm saying is that it can help to provide world peace. How can it do this?

First, all personal information—including financial transactions and voting—would be decentralized, anonymous, and ungovernable. This would give every person on the planet personal and financial freedom.

For instance, all personal transactions, from buying and selling a home to transferring health records, would happen seamlessly at minimal cost. These transactions alone will not influence world peace, but they could play an important role in way that I'll describe later in this book.

Secondly, there would no longer be voter fraud. A blockchain voting system will ensure the vote of all a country's citizens would be counted and counted correctly. Remember Joseph Stalin said, "Those who vote decide nothing. Those who count the vote decide everything." In a blockchain voting system, dictators, tyrants, warmongers, and incompetents can theoretically be voted out of office. With my proposed methods of providing monetary help for the poor and destitute people of the world, it's possible for them to collectively influence who runs their country. In this sense blockchain technology can potentially be an effective tool in providing world peace, if tied to the hydrogen revolution. As you will see in the last chapter of this book, today's version of blockchain technology is too difficult to use, is subject to cyber-attack, and lacks the ability to track fraudulent activities. These problems can be overcome in the future. It's just a matter of identifying a problem and solving it.

So, let's get back to why the worldwide hydrogen revolution is taking place now. During my later years as an engineer, I was responsible for the design and development of transport refrigeration equipment for the Thermo King Corporation. As you probably know Thermo King is a well-known name that is seen throughout the world on refrigerated trucks and semitrucks. In my engineering management capacity, I became interested in lithium batteries and hydrogen fuel cells since either could be used as a replacement for the diesel engines used with Thermo King equipment. Back then (more than twenty years ago), I concluded that lithium batteries and hydrogen fuel cells were much too expensive. Besides using hydrogen as a fuel still conjured up negative images of the Hindenburg disaster for many people. Believe it or not, the Hindenburg disaster was a demonstration of hydrogen safety as I'll explain later.

But times have changed. Elon Musk has changed the landscape with his lithium battery-powered electric passenger cars, and a company called Plug Inc. has proven the safety and reliability of hydrogen fuel cells. By mass producing lithium batteries in a robotically operated gigafactory, Musk has been able to dramatically reduce the cost of lithium batteries. During this same timeframe, Plug Inc. has produced more than fifty thousand hydrogen fuel-cell powered forklifts that have a remarkable safety and reliability record. While the cost of hydrogen fuel cells is still quite high relative to lithium batteries, Plug has built, and is currently operating, a robotically operated gigafactory which will dramatically reduce fuel cell and hydrogen fuel costs.

Needless to say, vehicle manufactures—including Thermo King—and governments are closely following these cost reduction and safety developments. While some are acting in response to fears of climate change, the economic and geopolitical aspects are also coming to the forefront. What if hydrogen fuel cell applications become less expensive than those of fossil fuels? What if every country in the world could become energy independent and didn't have to rely upon countries like Saudi Arabia, Iran, or Russia for oil and natural gas? How about rising gasoline prices resulting from oil wells reaching their peak output.

Then of course there are climate-change advocates who sincerely believe the world as we know it will be destroyed by a rising ocean levels. In fact, the most outspoken advocates think this destruction will occur in only ten or fifteen years if we don't act quickly. It then goes without saying that hydrogen derived from water and renewable or nuclear energy is an obvious answer.

Interestingly these same advocates, are the ones who shut down the nuclear power industry years ago but are now advocating nuclear power plants since they don't produce the greenhouse gas—carbon dioxide.

The fear of climate change—whether real or imagined—will be mainly responsible for accelerating the hydrogen revolution. However, many countries are tired of relying on other countries for their increasing demand for oil and natural gas. The idea of becoming energy independent is appealing. In fact, much of Europe's energy supply comes from Russia, and their counteracting hydrogen infrastructure is well on the way. Because of having almost no domestically produced oil, natural gas, or coal, Japan appears to be going "all in" for hydrogen. Japan's "all in" hydrogen obsession is probably a result of their abandonment of nuclear power because of the Fukushima nuclear plant disaster. While China has a lot of domestically produced high-sulfur coal, it is currently the world's largest importer of oil and natural gas. In fact, because of a lack of smokestack emission controls, China has five of the world's top ten cities with the most air pollution. However, because of its large population and economic growth, China is currently building a new coal-fired power plant almost every week. Needless to say, hydrogen made from renewable (and possibly nuclear) sources has become their primary energy focus. While not much was publicized in the United States about hydrogen being part of the Summer Olympics in Japan and the Winter Olympics in Beijing, both countries emphasized the use of hydrogen. China utilized 212 hydrogen fuel cell powered busses to shuttle Olympic contestants.

Even Saudi Arabia is going full bore in its hydrogen developments. Do they know something we don't? Are their oil wells reaching a peak level of production? Then there's India, with its large population and land area. Are they adopting hydrogen because they fear climate change, or do they just want to become energy independent? Moreover, from what I'm hearing, Australia appears to be gearing up to become a world supplier of liquefied hydrogen because of their vast amount of unused land and plenty of sunlight that could be used to produce hydrogen from water using electrolysis (a process of splitting water into its atomic parts: hydrogen and oxygen).

While the rest of the world is shifting to hydrogen, the United States is apparently going "all in" for battery-powered electric cars by planning to build a half million battery charging stations throughout the country. As a result, there appears to be less emphasis on building a hydrogen fueling infrastructure. Our leaders appear to ignore the fact that electrically powered vehicles larger than passenger cars must eventually use hydrogen fuel cells because the number of batteries they need are too heavy and consume too much cargo space.

Although battery-powered cars have caught on in the United States, hydrogen-powered electric fuel cell cars and trucks, busses, trains, ships, and aircraft have not. I would even venture to say, that most people in the United States don't have a clue about hydrogen fuel cells and what's happening in this regard around the world. But don't despair, a lot of things are happening behind the scenes. As previously mentioned, a company named Plug, and many others, have seen the handwriting on the wall and are paving the way.

That brings me back to why the hydrogen revolution can lead to a future of world peace and prosperity. It begins with the idea that since hydrogen fuel can be made from water and renewable and/or nuclear energy it could eventually make every country in the world energy independent. The form in which this process takes place could be a real game changer if the world's poor and destitute people were to become hydrogen fuel suppliers. Stay tuned while I show you how and why this seemingly impossible idea might be possible. I guarantee it will stretch your imagination.

CHAPTER 1

THE ELECTRIC PASSENGER CAR

I've chosen the battery-powered electric passenger car to be the first chapter of this book because, believe it or not, it represents a starting point for the hydrogen revolution. Although the battery electric passenger car has recently become a focus of interest to many people, it's still not fully understood. As a byproduct of this recent focus more people will begin looking for alternatives to using fossil fuels like gasoline. In other words, the electric vehicle (EV) has become an increasingly popular topic of conversation as advocates and skeptics discuss the pros and cons.

The US government is currently planning to spend taxpayer money to install half a million EV battery charging stations throughout the United States. In this regard, there's good news and bad news. The good news is that EV's serve as a platform for hydrogen fuel cells that can be used as an alternative to batteries. In fact, many of the costly repairs to internal combustion powered cars will be dramatically reduced by using EV's, which have fewer and more reliable parts. The bad news is that emphasizing batteries over hydrogen fuel cells only results in delaying the inevitable. The inevitable being widespread use of hydrogen as the portable fuel of the future.

Unfortunately, most people in the United States today are not very well informed about hydrogen fuel, and that hydrogen fuel cells are an alternative to batteries. Adding to this misunderstanding is a belief that hydrogen fuel is too dangerous. When people think of hydrogen, they tend to think of the Hindenburg disaster. Well, be that as it may, let's see how the battery-powered passenger car might play out in the long run.

Did you know that electric cars date back to the early 1900s. However, with the advent of the internal combustion engine and cheap gasoline they quickly lost favor. Because of America's dependence on foreign oil and gas shortages in the early 1970s, electric cars again began to receive serious consideration. Yet, the electric cars that were produced at that time suffered from limited performance—usually topping out at speeds of 45 miles per hour—and limited range—about 40 miles before needing to be recharged. Needless to say, with the advent of the lithium-ion battery and improvements in motors and electronics, electric cars are now making a comeback, thanks in large part to Elon Musk and Tesla. But is this third comeback sustainable?

Despite their recent surge in popularity, battery-powered electric passenger cars are not without problems. Thanks to the vision and innovativeness of Mr. Musk and his engineers, many of these problems have been, and continue to be, solved. For example, in 2010 lithium batteries had a cost of about $1,000 per kilowatt. They are now available for about $150 per kilowatt due to mass production. Some forecast the cost to drop to $70 per kilowatt, which would make the cost of the 100-kilowatt Tesla Model S battery-powered passenger car cost competitive with cars using internal combustion engines. Realizing that owners of battery electric cars will face battery replacement costs after 300 to 500 recharging cycles (about 125 thousand miles), current Tesla cars are currently sold with an eight-year battery replacement guarantee and a clever 90-minute battery replacement capability.

Even though there is a considerable fuel cost advantage for battery electric cars, there are still some disadvantages that can't be addressed with today's technology. First, there is the issue of recharging. Fast over-the-road charging takes more time than filling a gasoline fuel tank or a hydrogen fuel tank, and fast charging takes a toll on battery life and performance. The life expectancy of lithium-ion batteries is much shorter than gasoline-powered internal combustion engines or fuel cells. The cost to replace the batteries (which is paid either in the initial price of the car with a replacement guarantee or as an end-of-life direct purchase) offsets the fuel savings advantage in the long run. Because increasing the battery's kilowatt size means adding considerable weight to the vehicle, the maximum range tends to be limited at some point (probably the TESLA Model S with its beginning of life 400-mile-range). Moreover, lithium batteries have been shown to be susceptible to thermal runaway, leading to fires and explosions. These explosions have caused significant injuries and loss of property around the world. Then there is the issue of cobalt. Cobalt is an expensive ingredient used in the production of Tesla-style batteries and is almost singularly supplied by mines in the Congo region of Africa. Without competing sources, a major supply-and-demand issue could arise as the number of battery electric vehicles increases worldwide.

Lastly, even though many people believe they are reducing greenhouse gas emissions by buying EVs, the facts may surprise you. Current studies show that the electricity needed to recharge these batteries is mostly generated with coal and natural gas fired power plants. It turns out that the total greenhouse gas emissions are not much less for battery electric vehicles than for gasoline-powered internal combustion engine vehicles. Although batteries discharge electricity at a very high efficiency, their electricity is derived from power plants that have very low efficiencies. Compounding the problem, studies show greenhouse gas emissions associated with mining and recycling lithium batteries is considerably more than that generated by the production of internal combustion engines.

To be more specific, a typical EV battery weighs one thousand pounds, about the size of a travel trunk. It contains 25 pounds of lithium; 60 pounds of nickel; 44 pounds of manganese; 30 pounds cobalt; 200 pounds of copper; and 400 pounds of aluminum, steel, and plastic. Inside are over six thousand individual lithium-ion cells. Generally speaking, these materials come from mining. For instance, to manufacture an EV auto battery, you need to process about 25 thousand pounds of brine for the lithium, 30 thousand pounds of ore for the cobalt, 5 thousand pounds of ore for the nickel, and 25 thousand pounds of ore for copper. All told, you need to dig up almost 500 thousand pounds of the earth's crust to make one EV battery. Moreover, about sixty-eight percent of the world's cobalt, a significant part of a lithium battery, comes from the Congo. Their mines have no pollution controls, and they employ children who die from handling this toxic material.

That's just to produce the batteries. The end of life of lithium EV batteries is another story. While a typical lead acid car battery can be economically recycled, lithium batteries cannot be economically recycled at this time. Recycling requires very expensive processes and a large consumption of energy. In addition, much of the processing requires dealing with toxic chemicals.

In contrast only a fraction of the amount of the earth's crust is mined—mostly platinum to coat fuel cell electrodes—to make an equivalent automotive-sized hydrogen fuel cell. Moreover, hydrogen fuel cells do not contain toxic materials, and 95 percent of the platinum is recoverable in a recycling process. In fact, Ballard, Inc. (a fuel cell company), claims it will provide a 30 percent trade-in discount toward a replacement.

What about vehicles larger than passenger cars? Well, as previously discussed battery weight and space becomes even more of an issue with larger vehicles like semitrucks. Because of their much larger size, batteries just can't compete with gasoline and diesel engines. But hydrogen fuel cells can. In other words, if semitrucks are battery powered, a significant amount of potential cargo space and weight limitations would be displaced by batteries to retain the same range before refueling. Hydrogen fuel cells solve that problem.

How about hydrogen being the portable fuel of the future for all vehicles: trains, ships, and aircraft? Studies show that we haven't more than forty more years supply of gasoline or diesel fuel at current usage rates. Whether or not you believe that is not the issue here. The real issue is economics. Take note of how the battery-powered electric vehicles evolved. Although they have many good attributes, electric cars are now on their third trial and still have problems with cost, weight, battery life expectancy, range reduction over time, recharge time, material supply issues, loss of range during cold weather, safety, and recycling. Could it be that hydrogen fuel cells will replace or supplement the lithium battery for passenger cars? Could it be that innovation, new technologies, mass production, and free enterprise will solve fuel cells current cost and fuel supply problems?

Recently Elon Musk referred to hydrogen fuel cells as "fool" cells. And why not? His current venture into battery electric cars has made him one of the wealthiest people on the planet. Being an innovative and forward-thinking person, he can see the competition that hydrogen fuel cells pose to his current automobile empire. However, I have good news for Elon. In my humble opinion, hydrogen fuel cells won't replace batteries but will eventually be used as range extenders. In other words, as the price of fuel cells drop due to mass production and an over-the-road hydrogen fueling infrastructure is put in place, a battery-fuel cell hybrid combo will become a viable solution. This is because batteries are more efficient (from a fuel consumption standpoint) than fuel cells. In which case an optimum-sized battery can power a vehicle most of the time while an optimum-sized fuel cell and energy from braking can be used to recharge the battery. When additional power is required, such as for acceleration and going up hills, the fuel cell can kick in to supply the additional power. Needless to say, with such a combo over-the-road refueling with hydrogen will take much less time, and will cause the previously mentioned half million over-the-road battery charging stations to become obsolete.

While fuel cells alone have a considerable weight advantage, they don't have much of an occupied space advantage—only in passenger cars—due to the cylindrical shape of their pressurized hydrogen fuel tanks. In addition, the more efficient battery operation (the energy source to wheel energy efficiency is about 80% for batteries vs about 52% for fuel cells—and about 25% for internal combustion gasoline engines) should improve fuel cost, and the compact shape of batteries should offset the cargo/passenger space issue. Moreover, by locating a flat battery pack array underneath the vehicle cargo/passenger space, the resulting low center of gravity will add to stability. One other point is that batteries provide faster acceleration than fuel cells.

As you can see, the same innovative ideas that drove our recent battery electric car revolution can also be applied to hydrogen fuel cell vehicles. The question is when is it going to happen? To answer this question, we need to look at what is happening now, and ask are there other driving forces that will accelerate the process? The answer is yes. There are other driving forces that will accelerate the process.

CHAPTER 2

HYDROGEN ECONOMICS

Technology changes take place all the time, and it's driven mostly because of economics and/or demand. Not too many years ago there were no internet or cell phones. Today people can't live without them even though they are an added expense. Will hydrogen fuel be the next major technology shift? How about countries around the world adopting hydrogen to make them energy independent? How about a fossil fuel supply shortage? How about a solution to climate change? How about making money?

Let's begin by focusing on countries around the world wanting to become energy independent. Since hydrogen can be extracted from water (by electrolysis and other yet to be commercially proven methods) it's obvious that if it can be done economically without the need for fossil fuels, it will be done. Believe it or not, solar energy has now become competitive with fossil-fuel fired electric power plants and may be nearing an economic tipping point. Also, since low-cost mass production of hydrogen electrolyzers and hydrogen fuel cells are fast becoming a reality, let's explore what this could mean.

There are currently about 290 million passenger cars in the United States and about 120,000 gas stations. My calculations show that if we fueled with hydrogen, we would need 380 million kilowatts of solar-derived electric energy to fuel 276 million cars per year (or 95% or all passenger cars). If the hydrogen fuel is dispensed at the 120,000 gas stations, my estimated total investment over thirty years is $7.2 trillion ($240 billion invested per year if evenly applied). As a result, each of the 120,000 gas stations would operate at an annual gross income loss (that's right, I said loss) of $248,000 per year after paying a thirty-year amortized loan at 3 percent. This number assumes no government tax subsidy; a government tax of 30 percent on hydrogen sold at $8.00 per kilogram[1]; the cost of all associated processing, delivery, and dispensing equipment; a northern climate that

---

1 In terms of over-the-road miles per gallon, $8.00 per kilogram of hydrogen is equivalent to about $3.50 per gallon of gasoline.

receives 15 percent less sunlight; and today's current 20 percent solar panel efficiency. It also assumes 60 kilowatt hours of electricity to convert water into one kilogram of compressed hydrogen (the energy equivalent of one gallon of gasoline). With this in mind, does the above analysis and assumptions preclude my hydrogen revolution prediction? Let's see.

Obviously, what I've just proposed is not a workable solution. However, it does identify how hydrogen fuel, derived from solar energy, can become economically feasible (i.e., at $8.00 per kilogram). The first thing to consider is lower cost government loans—perhaps 1 percent rather than 3 percent—and reducing the 30 percent government tax to about 20 percent. These actions alone will result in an annual gross income gain of $1.0 million per gas station per year (that's right, I said gain not loss). Now let's assume a solar panel efficiency of 26 percent rather than 20 percent. This efficiency improvement has been demonstrated in laboratories and will most likely be commercially available in the next few years. The annual gross income gain per gas station with this solar panel efficiency improvement would be $2.1 million.

Let's go one step further. As you have probably noticed, I've assumed compressed hydrogen fuel that's derived from 60 kilowatt hours of electricity. However,

If the hydrogen is centrally produced and delivered to each gas station, it will most likely be in a liquefied (cryogenic/frozen) form. This will add another 10 kilowatt hours to the process and will reduce the gas station's gross income to about $496,000 thousand per year rather than the previously mentioned $2.1 million. Assuming four employees at $50,000 per year, each gas station owner will then gross $296,000 per year before other expenses. Since the average gas station owner currently has a net income of about $85 thousand per year, this idea still looks pretty good.

Although centrally producing liquefied hydrogen and trucking it in is the most likely scenario, I have another idea. For gas stations located near farmland, the liquification process could be avoided. Just think of farmers that might have many acres of farmland, some of which could be used to generate and deliver compressed hydrogen to gas stations that might be as far away as fifty miles. After all, current United States farmland crops net about $30,000 per 100 acres per year, while the addition of solar-produced hydrogen might significantly increase that amount.

If a farmer with 100 acres of farmland wanted to devote just 10 acres to solar hydrogen production, and if the farmer got a 15-year, 1-percent government loan to pay for solar panels with a 26 percent efficiency, then the farmer could generate a gross profit of more than $165,000 per year after a 20 percent government tax. In this scenario the farmer delivers compressed hydrogen to designated gas stations in a tanker trailer for $6.00 per kilogram. (Whether or not the trailer's contents are transferred to a gas station storage tank or is dropped off in propane-like bottles, needs further economic study.)

I'm going to briefly interrupt the description of my idea at this juncture to make two additional points that I believe will help to offset possible criticism about what I'm advocating. The first point is that of the above-mentioned gross profit of $165,000 per year a portion of that money would need to go towards adding additional solar panels to offset an approximate 0.5 percent of lost solar panel efficiency per year, which will also add to the amount of land being used. The second point concerns solar panel life expectancy and recycling. Although manufactures may state a 20-year expected solar panel life, in reality that life can be extended to 40 or 50 years with proper maintenance. Of course, because of a continued loss of efficiency, the panels would most likely reach a point at which replacement and recycling would make the most economic sense.

Now, getting back to the idea that I'm trying to put forth, let's see how the gas station owner makes out by selling the compressed hydrogen for $8.00 per kilogram. If the gas station owner sells about 1,200 kilograms of hydrogen per day (supplied by three farmers using thirty acres of land—plus additional acreage over time as more solar panels are added—at $8.00 per kilogram, he or she would have a gross income of about $700,000 per year. This is, of course, is a better deal than the above mentioned centrally supplied liquefied hydrogen and may eventually be the best arrangement even for gas stations located in congested cities. Besides, the farmer suppliers could be widely distributed individuals rather than big oil companies with centrally located supply depots.

The idea of thousands of farmers increasing their income appeals to me and forms the basis for what I'll be presenting later in the book. As I see it, in this scenario, the big oil companies could adapt. The top five US oil companies have an annual revenue of about $1 trillion. Since their new business model would consist of supplying oil for pharmaceuticals, synthetic materials, plastics, lubricants, cosmetics, and asphalt, it would result in lost jobs. There would be, however, an equivalent increase in jobs from the solar/hydrogen side of the equation. Besides, as previously mentioned, the amount of oil appears to have a limit—maybe forty years at our current rate of use—so why use it all up when it could serve so many other useful purposes. The key words here are "current rate of use." As you will see. in a later discussion, the current worldwide usage rate is increasing significantly.

Also, considering the farmer supplier scenario, the idea of mobile fuel delivery might make sense. Why not include fuel deliveries to highway rest areas and predesignated parking lots? Better yet, with today's satellite communications systems, a car or truck driver could call ahead when their fuel tank is approaching empty. Moreover, with the farmer making the consumer point of sale, the sale price could be $8.00 per kilogram rather than $6.00.

Let's further analyze what I've just said about the farmer supplier idea. About 380 million kilowatts of solar-generated electric power would be required to supply 95 percent of electric passenger cars' energy. This is about 80 percent of the US electric power being generated today. But here is the good news, it will initially require only 15 million acres of farmland. With about 920 million acres of farmland in the United States, less than 2 percent of all farmland will be required to generate the required solar energy. The point being that if all fossil fuel usage from cars and trucks to aircraft and home heating is replaced by solar-generated hydrogen, it's not going to make a significant impact on the United States food supply.

Here is another interesting point. If all passenger cars, trucks, and other vehicles, were powered by hydrogen fuel cells, they could replace almost all the electric power generated by the United States. In other words, by plugging these vehicles into an electric supply system, you will be able to continue with electric power in the event of a power plant blackout. More importantly, the United States could be protected in the event of a nationwide cyberattack on our centralized power grid.[2] In contrast, since battery-powered electric vehicles would be mostly dependent upon the electric grid, a recharge would not be possible. On the other hand, a hydrogen fuel supply infrastructure would provide continuous support while the grid is being restored. Some estimates are that in the case of major grid disruptions, restoration could take up to eighteen months.

Before leaving this chapter, I need to say something about what's happening today in the United States and around the world. As previously mentioned,

---

2   The US currently has 55 interconnected power transfer stations that form a nationwide centralized electric grid. If just 14 of these stations were to experience a cyberattack from a rouge group, or country the entire grid system could be shut down.

a company named Plug, Inc., has sold more than 50,000 hydrogen fuel cell retrofit kits for forklifts. Their base companies are Amazon and Walmart. As a result, Plug has installed liquefied hydrogen refueling stations all over the United States. Interestingly, Plug's selling point was economics in spite of higher fuel cell cost relative to batteries and the high cost of hydrogen fuel. Since forklifts are generally used around the clock, they need to stop for extended periods of time for battery recharging whereas fuel cells can be refueled in minutes. When considering downtime, space for recharging, and battery disposal, the fuel cell solution became the least expensive choice. Now Plug has built a gigafactory to cut fuel cell prices in half. Additionally, they are now robotically producing fuel cell operated electrolyzers in this same factory. Moreover, Plug is actively helping to create publicly available hydrogen fueling stations throughout the US and in other countries. So, what does this mean for the hydrogen revolution? Well, first, straight trucks (a truck in which all the axles are attached to a single frame and not 18-wheelers) and busses generally return to a home base for refueling. In which case localized hydrogen dispensers can service the straight truck or bus fleet. However, with increasing numbers of warehouse, bus, and truck depots hydrogen supply locations being installed all over the United States, it makes sense to make them available to the public if there is a profit motive. This would help solve the hydrogen infrastructure problem by making hydrogen refueling available to buyers of hydrogen fuel cell vehicles. Additionally, with governments and private industries sponsoring hydrogen highways throughout the world, the transition to hydrogen fuel has already begun, and car, bus, train and truck manufactures are following

CHAPTER 3

THE HYDROGEN GOLD RUSH

In my opinion straight trucks and semitrucks will be the primary driving force in the hydrogen revolution in the United States. My gold rush idea begins with about four thousand semitruck refueling stops across the nation employing natural gas to hydrogen reformers. Why would entrepreneurs build so many refueling stops? Because of the enormous profits that can be obtained. Natural gas reformed hydrogen currently costs about $4.00 per equivalent gasoline gallon (one kilogram of hydrogen). If hydrogen is sold for $8.00 per kilogram, there will be an enormous profit incentive because the equivalent miles per gallon for an electric hydrogen fuel cell powered semitruck (and other vehicles as well) will be more than double that of gasoline-powered vehicles. If the price of diesel fuel is

$3.50 per gallon, then $8.00 per kilogram for hydrogen would be offsetting.

The $4.00 per gallon profit would be like discovering gold. Even though half of the $4.00 profit might go to the government, compression, storage, and dispensing costs, the net profit to the truck stop owner would still be significant. Therefore, both the government and the truck stop owner would be strongly incentivized to get the process moving. It would even benefit the government to provide strong cost incentives to semitruck manufacturers to build hydrogen fuel cell trucks and to manufacturing companies to build thousands of low-cost natural gas reformers rather than thousands of fast charging stations for battery powered EV's. Truck stop owners would purchase these reformers because of their profit potential. Sort of like a potential gold miner buying gold mining equipment before mining the gold.

Let's look at a company named HyGear that can provide truck stops with 1,000 kilogram per day natural gas to hydrogen reformers. Their reformers take the space of three forty-foot-long containers that should easily fit on most truck stop parking areas. If the mass-produced cost is $500,000 for each reformer system, amortized at 3 percent over ten years, the yearly cost would $58,200 per year. Are you getting the picture? A thousand kilograms per day at $2.00 per kilogram would net the truck stop owner $720,000 per year gross profit. Since most semitruck refills will be on the order of 50 kilograms, a reformer system could fill twenty semitrucks per day. But the gold mine gets even better. Twenty semitruck refills is a small number compared to the number of refills typically required per day when the number of hydrogen-fueled trucks and semitrucks increase over time.

So why not continue with hydrogen derived from reformed natural gas rather than switch to the higher cost of solar generated hydrogen? This is a good and logical question, and here is a simple answer. Besides the fact that the hydrogen producing equipment footprint will eventually exceed available truck stop space, there isn't enough natural gas in the United States to supply hydrogen to all vehicles (cars, trucks, trains, ships, and aircraft) with reformed hydrogen for a long period of time. According to the US Energy Information Administration (EIA) there is about 2,320 trillion cubic feet of known technically recoverable natural gas in the United States. If we continue our current usage of about 28 trillion cubic feet per year, we have enough natural gas to last another eighty-three years. But what if all transportation vehicles were to begin using reformed natural gas and electric power plants were to steadily increase their natural gas usage as well? As I see it, we could more than triple our usage. In which case we would have only twenty-three years' worth. That's not a good idea. While the gold rush scenario sounds great at first, it's only a temporary solution to getting the hydrogen revolution started. As with any commodity, it's a question of supply and demand. If natural gas prices or shortages reach a breaking point, a renewable energy or nuclear option will come into its own—especially as the number of hydrogen-fueled vehicles increase. That's not to say that government legislation will intervene before supply and demand factors take hold.

Another option is to invest about $4 billion to place small scale natural gas to hydrogen reformers at four thousand major highway truck stops across the country. It would be like adding an air pump, vacuum device, or vending machine to the gas station property. These reformers could service semitrucks or passenger cars and get the ball rolling so to speak. At $8.00 per kilogram of hydrogen the gas station owner could be incentivized because of a share in the potential profits. By doing this in advance of significant hydrogen fuel cell vehicles on the road, it would clear the way for those who wouldn't buy them otherwise. Besides, the government or a billionaire could do this without blinking an eye. Even though profits probably won't be realized for two or three years, the potential long-term return on investment could be significant.

In any event, smaller gas station owners will most likely see the profits from natural gas reformers and begin mining their own gold, so to speak. A case in point is a 100-kilogram-per-day reformer can fit on a footprint the size of an automobile parking space. The gas station owner pays $100,000 to install the equipment and then pays a 3 percent amortized amount, or about $11,600 per year for ten years. But at $8.00 per kilogram sold, the owner would pocket $280,000 per year. What a deal! You can almost see the camel's nose is getting under the tent and causing the hydrogen revolution to accelerate.

CHAPTER 4

MY WAY

F rank Sinatra made the song "My Way" famous in the later years of his life. It's one of my all-time favorites. The song tells a story that reflects upon a long life lived. Looking back on my eighty-year life, I've been blessed with a few talents and deprived of many others. I've encountered good fortune and bad fortune, good times and bad times, happy times and sad times. In fact, I've come to believe there is no such thing as coincidence. Not only that, I also believe that our human life is a trial that will be evaluated at the end. As I see it, life is a series of trials that require a series of responses. How we respond to these trials will probably determine our afterlife assignment.

So, please allow me – at this point in the book—to tell you some things about myself that have shaped my life, and why I've chosen to do things **my way**. With regard to my God-given talent, it became a question of recognizing what it is and trying not to squander it. It's an ability I seem to have that I can't explain. While some people are good at languages and others can play musical instruments—I'm not good at either—however, I am good at thinking creatively. After reading this book you can decide whether or not that is true. My creativity comes out in many ways, which might explain why I've written this book, and others, during my retirement years. While some might think my creative thinking belongs in fantasyland, others may not. In any event, with the books I've written, I will have left my thoughts and ideas for the next generation. Hopefully my ideas will have a positive influence as our future unfolds.

So, here is a brief summary of my life story and why I've done things the way I have—**my way**. As a child I was interested in piecing puzzles together, figuring out how magic tricks work, and solving riddles. Not surprisingly this attribute has followed me to this day. Perhaps it was this attribute that led me to become an engineer. Not only was I interested in knowing how things worked, I was also interested in inventing new things or creating ways that would make things work better. My unconventional way of thinking has made me somewhat of a maverick. I question everything, especially consensus thinking. I've never been a person who goes along to get along. Putting it another way, I have not, and will not, follow the crowd. Perhaps as you read this book, and others I've written, you will notice that I have certainly not been bound by conventional thinking.

Growing up in a poor family, in a poor neighborhood just outside of Boston, I occasionally saw the things rich people were able to do with their money, and I wanted the same. When I was in middle school, one of my classmates, Wayne Libby, told me that he was going to go to college and become an engineer. My first response was what is an engineer? After he told me what an engineer was, I asked why he wanted to become one. To make lots of money and become rich, he told me. At that point he had my attention. It had nothing to do with Wayne's interest in technology and science. On the other hand, I had always been interested in technology and science, and the idea that I could get rich by doing things I liked was very appealing.

That evening I told my mother and father I wanted to go to college and become an engineer. My older brother (ten years older than me) was apprenticing to become a (now obsolete) photo engraver. It was going to take five years for him to learn the trade. I think my family was disappointed that I wouldn't pursue a trade too. After all, the only person in my family who had gone to college was my rich uncle Norman. Besides, where was I going to get the money to go to college since my family certainly didn't have it?

In my neighborhood everyone was expected to finish high school and get a job doing something. I observed that my uncle Norman, who had gone to college, owned his own business and appeared to be doing quite well. My family and my friends' families, who hadn't gone to college, all worked in various jobs and lived in a poor neighborhood. Putting two and two together, I came up with four. But the question remained: how could I go to college if my family couldn't afford to pay my tuition?

Being optimistic and undeterred, I chose to go to Boston Technical High School, where I could learn technology and science and get ready for pursuing an engineering degree. In my senior year I was accepted at Northeastern University, which had the largest cooperative education program in the country. It was a five-year bachelor's degree program. Students would work ten weeks and go to school ten weeks. In this way they could earn enough money to pay for their tuition, with a side benefit of getting hands-on business and engineering experiences. The problem was that the co-op program didn't begin until the second year, which meant that I had to pay the first year's tuition. When my uncle Norman heard about my financial problem, he came to the rescue and paid it for me. Thank God for my rich uncle, Norman!

In my second year at Northeastern, I joined the Reserve Officers Training Corps (ROTC) program and was obligated to two years of active duty in the US Army. When I did graduate with a degree in mechanical engineering in 1964, I immediately entered the Army as a second lieutenant. Fortunately—or unfortunately depending upon how you look at it—my active-duty service included a tour of duty in Vietnam. As you might surmise, my experience in Vietnam has had a profound effect on me. For one thing, it strengthened my Christian beliefs. As the old saying goes "there are no atheists in foxholes."

While my service in Vietnam didn't make me rich, I was able save almost all my income—which included a military bonus called combat pay. This enabled my wife and I to purchase a house after my active tour was over, and I had an engineering job with Westinghouse. While my work on advanced nuclear steam generators was interesting, it didn't make enough money to make me rich—as Wayne Libby had once thought. On the other hand, I did notice some Westinghouse managers who seemed to be doing quite well. The question then was how could I become a Westinghouse manager? To make a long story short, I read a book called *How to Win Friends and Influence People* by Dale Carnegie. Following his good advice, and some other fortunate circumstances, I became an engineering manager. My salary was very good, but it still didn't make me rich.

My purpose for being rich became quite different at a later point in my life. As time went by, I didn't want money so that I could indulge myself; rather I became more interested in how having money might help other people? After witnessing firsthand, the extreme poverty in Vietnam, I developed a desire to someday help. Part of the reason for writing this book is to find a way to help impoverished people around the globe. I don't believe we can help poor and destitute people by giving them a handout. Like the old saying goes, if you give people a fish, you feed them for a day. If you give them a fishing rod, you feed them for a lifetime. In this book I'm going to share some "fishing rod" ideas that might ultimately enhance the lives of poor and destitute people around the world for a lifetime.

As President Ronald Reagan once said, "You can't enrich the impoverished by impoverishing the rich." No truer statement has ever been spoken. I say thank

God for the rich. While a poor person may look at a rich person with envy and contempt, the fact is, without rich people everyone would be impoverished. I mean really impoverished! The poor in the United States are lucky compared to poor people in other parts of the world. In the United States there is a safety net for the poor that is mostly paid for –by the rich. (The top 10 percent of taxpayers pay about 70 percent of the US tax revenue.) Moreover, rich people provide jobs either through owning a business that employs people or through buying things like luxury homes, cars, planes, and yachts. Best of all, rich people provide an incentive for poor people to become rich.

To me the greatest problems facing planet Earth today are poverty, hunger, and disease—not an exponentially increasing population, nuclear war, lack of non-renewable resources, or climate change. If we could eliminate the poverty problem, I believe the potential problems of overpopulation, nuclear war, and other legitimate concerns would be dramatically reduced or eliminated. Just imagine if the entire population of the world had the economic and political freedoms enjoyed by the citizens of the United States. Even the poorest of the poor in the United States have access to emergency medical services and a government-sponsored safety net for those who are incapable of taking care of themselves. Although there is a wide range of income disparity, those at the top of the ladder directly and indirectly help those at the bottom. And because it's possible for some on the bottom rung of the ladder to climb to the top, the rich provide an incentive of hope. It's this incentive that helps to grow an economy, and when an economy grows it provides jobs and income for those who would like to achieve a better standard of living.

As the saying goes a rising tide lifts all boats. The United States, while not perfect, stands as the current best example of what humans can accomplish. In contrast, most of the world's population is living in miserable conditions, without much chance of improvement. If we can provide them a chance, if we can give them hope, I believe the human spirit will rise to the occasion. Perhaps they are craving the freedom and prosperity that they see every day. With television broadcasts, movies, and the internet, the world is being exposed to how a better life might be, and I believe their inner being wants it. In my opinion, when they have it, they will choose to have fewer children, and they won't accept a repressive government whose objective is to take away the spirit of freedom. With more and more people demanding electric power, automobiles, modern appliances, and other goods and services that make for a better way of life, the world economy—including the economy of the United States—will grow and prosper.

With about 35 percent of the world's population currently consuming most the world's non-renewable natural resources, it's logical to assume that when, and if, the rest of the world catches up, there may not be enough to go around—especially when most of these resources are currently being thrown away in landfills.

To my way of thinking, having abundant and available energy, land, and water is a key to solving worldwide poverty, hunger, and despair; and solar-produced hydrogen from water can provide a catalyst for making it happen. However, the problem of depleting natural resources and an ever-increasing demand for food also needs to be addressed.

Therefore, let's take a journey into a future where all of the world's population participates in the freedoms, health care, security, and comparative wealth that most Americans experience today. But in order for this to happen we need to begin by thinking out of the box.

CHAPTER 5

OUTSIDE OF THE BOX

The first three chapters, and the introduction of this book, were a summary of how and why the hydrogen revolution can and will happen. In these chapters two new ideas were proposed. The first idea was called the new gold rush where nationwide truck stops create hydrogen from natural gas for $4.00 per kilogram and sell it for $8.00 per kilogram. The second idea was solar farmers who create, and sell, compressed hydrogen. Whether or not these ideas take hold is yet to be determined. But I'm not through. Since I consider myself to be an innovative thinker, the remainder of this book will be devoted to describing more new ideas. These ideas can not only drive the hydrogen revolution but can dramatically change the world in which we live. With this in mind, let me introduce you to another game changing, **"Out of the Box"** hydrogen-related idea.

This idea begins with Ocean Thermal Energy Conversion (OTEC). OTEC is the process of generating electricity using the temperature difference between the sun-warmed surface water and the cold deep water of the ocean. As a side benefit, a huge amount of salt water is converted into pure fresh water. An additional benefit is that deep water nutrients are brought to the surface, which results in improved fishing. Although these benefits sound great at first, the problem is that OTEC has a very low electric conversion efficiency—especially when surface waters are not very warm. However, I see a relationship between OTEC and nuclear power.

I know, I know, you may think that's crazy, but don't criticize me until you hear me out. Well in case you don't already know, before the well-known nuclear power plant disasters at Chernobyl and Fukushima, Westinghouse was building the largest crane system in the world. It was being constructed on a shoreline near Jacksonville, Florida, to aid in building floating nuclear power plants. Why? Because a floating nuclear plant cannot experience a meltdown. Why? Because water is used as the working fluid in current day nuclear plants. Why? Because the first nuclear plants were intended to be used in Navy ships. In Navy ships ocean water would quickly cause a shutdown of a nuclear reaction in the event of a wartime explosion. While nuclear-powered Navy ships have operated success-fully for decades, water-cooled nuclear power is not the best way to create nuclear fission electric energy. Since high temperature gas-cooled reactors, or sodi-um-cooled breeder reactors, can create almost as much nuclear fuel as they use, they were planned to be the next wave of nuclear electric power generation. But because of political pressure, they never got to see the light of day.

Let's resurrect the idea of nuclear power plants by placing modular (small scale) Navy nuclear reactors in an ocean environment near OTEC systems. Why? Because waste heat from these reactors can warm sur-face waters and make the OTEC system much more efficient and allow them to operate twenty-four hours per day. If we locate these OTEC/nuclear systems in the gulf streams along the East and West Coasts of the United States, the gulf stream waters will carry the heated water to the OTEC system.

With this novel idea a single 100-megawatt OTEC system with five 20-megawatt modular Navy nuclear reactor power systems will generate about 200 mega-watts of energy 24 hours per day. If we build one hundred OTEC systems (fifty along each US coastline), we could supply 2.5 billion kilograms of liquefied hydrogen each year. At an estimated production price of $6.00 per hydrogen kilogram, these one hundred OTECs could generate a gross profit of $15 billion per year and pay for themselves. Since my previous analysis showed that the United States could become hydrogen energy independent using solar-derived energy, the OTEC/nuclear systems would generate liquefied hydrogen fuel for export to other countries or as ocean fueling stops for large ships. Although hydrogen tanker ships could make liquefied hydrogen deliveries around the world, I recommend resurrecting the zeppelin (to be discussed in the next chapter).

Since the nuclear part of my OTEC proposal will generate 10,000 megawatts of nuclear electricity, we need to address a uranium fuel supply problem that may result. To address this problem, I recommend using sodium-cooled breeder reactors to not only supply the required long-term OTEC system fuel but to also generate more low-cost liquefied hydrogen. Because of a fear that some people have regarding the possibility of a "Chernobyl type" meltdown, I recommend modular breeder nuclear systems that can be located in very remote places. In fact, they could be part of a very large wind turbine structure. Let me explain.

Since the highest winds in the United States are along the eastern side of the Rocky Mountains, I recommend putting a wind turbine farm in the prairie land of North Dakota. You could fence off a sixty mile by sixty-mile square section with a five-mile vacant land perimeter, thus making a fifty mile by fifty mile location to place wind turbine-breeder nuclear power plant combos. In this remote area of the United States, you could build on-site mobile factories to produce 12-megawatt wind turbines (the largest wind turbines in use today are 5 megawatts) with some of them having 20-megawatt breeder nuclear power plants located in their pedestals. Five 650-foot-tall 40-foot diameter wind turbines (buried 75 feet deep) would be placed every square mile with one 20-megawatt breeder reactor-wind turbine combo located at the center of each square mile. Considering the large spacing between wind turbines, it's logical to also incorporate solar panels.[1]

Do you see how technological innovation can change everything? Well, that's not all folks. If large

---

1   For those readers who would like to see an illustration of my proposed OTEC system and wind turbine-breeder reactor idea, see my book *Reaching America's Destiny*. You will also discover why a 12-megawatt wind turbine is much more cost competitive than fossil-fuel-fired power plants but are currently too big to transport their blade parts from a factory to a distant location.

scale zeppelins can deliver liquefied hydrogen to any location in the world at a much lower cost than tanker ships or pipelines, I'll choose this form of transportation. Let me now explain why we need to resurrect the zeppelin.

CHAPTER 6

THE ZEPPELIN REVOLUTION

In the last chapter I recommended that zeppelins be used to transport liquefied hydrogen to any location in the world. My reasoning is that hydrogen is lightweight and, when in liquefied form, the containers can also be made of lightweight materials. As a result, mass-produced, low-cost, large zeppelins can be the least expensive (my estimate is $3 million per zeppelin if mass produced) means of transportation. Not only that, the zeppelin that I'm about to propose could be as revolutionary as hydrogen itself.

Let me begin my argument for zeppelins by telling you about a personal experience. Friends of my wife and I had a small airplane in which they took frequent weekend trips. We were invited to travel with them on several occasions and found the experiences to be quite enjoyable. We saw luxurious housing communities, interesting golf courses, farmlands and forested areas, the layout of towns and cities, and the intricacies of lakes and rivers that could only be seen by flying at a low altitude. Plus, we experienced the unique feeling of flying like a bird. What if we could all travel around the country in this manner for the price of a bus ticket? What if our flight had the accommodations, amenities, and spaciousness of a luxury cruise ship? What if we could board the craft with our car and be dropped off to explore areas of interest? What if it was less expensive than driving a car? How about using a large-scale zeppelin?

You probably know about the Goodyear and Met Life blimps, which have been replaced by drones, because of the televised aerial views that they provided during sporting events. The views in many cases were spectacular. Before sporting events, these blimps would fly over the surrounding area to show the layout of landmarks around the event area. Except for the take offs and landings of a high-flying jet aircraft the public is generally deprived of experiencing the beauty and intricacies of the landscapes below.

In the days of the zeppelins people were treated to wondrous views while traveling in the lifestyle generally provided by luxury cruise ships. For many people—especially retirees—getting from one place to another doesn't need to be fast. In fact, frequent stops or hovering at scenic wonders like the Grand Canyon could provide a bonus. If we could make low-cost zeppelins, they could become the preferred mode of travel to many who can't afford a plane—or even a bus—ticket.

The zeppelin I envision has a cargo lift capability of 1 million pounds at sea level, 700,000 thousand pounds at an altitude of 12,000 feet (high enough to cross most mountain ranges), and 140,000 pounds at 25,000 feet (high enough to travel in the jet stream). It uses liquefied hydrogen fuel to power fuel cells, which drive electric motors that turn propellers to control the zeppelin's position and flight direction. My proposed zeppelin could travel at speeds of more than 80 miles per hour and can stay aloft for a week before needing to refuel. When traveling in the jet stream, it could reach speeds of almost 200 miles per hour.

Because of cost and availability of the lighter-than-air gas needed to keep them aloft, most of my proposed zeppelins would be filled with hydrogen rather than helium. To deal with the controversial nature of hydrogen's flammability, the zeppelin's vertical takeoff, travel routes, and landings could be in unpopulated areas, and their first application could be for cargo only—no passengers, pilots, or crew. Some zeppelins, however, could eventually use helium for passenger travel.

To achieve the abovementioned attributes, my calculations show that my proposed zeppelin should be a torus, that is, about 500 feet in diameter and 200 feet tall. To keep their manufacturing cost low, they should be built in a factory that mass produces as many as 300 zeppelins per year. Before going into detail about why the mass-produced, low-cost zeppelin is potentially a game changer, I think it would be prudent at this point to provide some background information that will put my proposed zeppelin concept into perspective.

As you might have surmised, I find blimps and zeppelins to be quite intriguing. The word *blimp* generally means a "lighter than air" balloon-like structure made of a leak-proof fabric, whereas, zeppelins, like the Hindenburg, have a ridged structure with fabric on the outside and helium or hydrogen filled bladders inside. A compromise between the two would have a semi-ridged frame as part of the structure.

The Goodyear and Met Life blimps, which were used to provide aerial views of golf matches and other events, did not have a ridged frame. They maintained their shape and buoyancy by inflating and deflating airbags as the helium gas expands and contracts with altitude and temperature. At ground level, the airbags are filled. When in the air, the helium expands because the air is lighter at higher altitudes, and the air bags are deflated. For these blimps, using a non-ridged frame makes a lot of sense because of their relatively small size and marginal lift capability. However, the larger the blimp, the greater the lift capability. This happens because when you double the size of the blimp, you increase the volume of helium or hydrogen gas eight-fold, and an eightfold increase in volume is almost an eightfold increase in lifting capability. Therefore, a ridged frame can be used for larger sizes without much loss in the amount of cargo weight that it can lift.

Let's compare the Goodyear GZ-19 blimp to the Hindenburg. The GZ-19 is 192 feet long, 59.5 feet high, and 50 feet wide. The Hindenburg was 804 feet long, and 135 feet in diameter. The volume of the GZ-19 is about 300,000 cubic feet while the volume of the Hindenburg is over 7 million cubic feet—over 23 times that of the GZ-19—making a ridged frame more practical. By having a ridged frame, it could also lift a substantial amount of weight without damaging the balloon material. In the case of the Hindenburg, it was capable of lifting about 247,000 pounds in addition to its own weight.

My proposed zeppelin is much bigger than the Hindenburg! Utilizing the Hindenburg's net lift capability as a basis for calculating the proposed zeppelin's net lift capability, I estimate about 1 million pounds at sea level. My research shows that this is 600,000 pounds more lift capability than the largest zeppelin I could find on the internet. Yet my proposed zeppelin is 62 percent shorter than the Hindenburg and only 41 percent larger in diameter. As you can see, this diameter increase equates to about four times the lifting capability of the Hindenburg.

Let's pause here and look at two main reasons to reconsider zeppelins. First, there are new technologies that were not available when the zeppelins were discarded. Second, they would offer a low-cost alternative to current methods of transporting large objects.

So, what are the new technologies? Global positioning systems and advanced computer technology permit unmanned operation—thus permitting the use of low-cost hydrogen, rather than expensive and scarce helium, as the "lighter than air" gas. Kevlar, titanium, carbon composites, and aluminum alloys provide stronger and lighter frame materials. Finally, we can now use computerized design analysis, which is capable of evaluating wind forces and structures in a way that was not dreamed about in the days of the Hindenburg.

While mass production was available when zeppelins were discarded, there was no need to mass produce them in the numbers I propose. Additionally, robotic assembly techniques were not in use at that time. And, because of my proposed torus shape, it would not be necessary to provide each zeppelin with an expensive ground-based enclosure. Although a few service and maintenance silos will be needed, there is no need for each zeppelin to have a storage enclosure during severe weather. When the zeppelins are not in use, and during severe weather, the zeppelin can be anchored on the ground using an automated low-cost strapping system to brace the zeppelin during windy conditions and doesn't need to have its nose end facing into the wind. It can also fly above the weather by dropping off its cargo if necessary.

Besides transporting people, these giant zeppelins could carry massive amounts of cargo, which would normally be transported by train or semitrucks (about thirty semitruck containers worth at a fraction of the cost), thus reducing number of semitrucks on highways. Have you ever wondered why the pre-manufactured house has never caught on? It's probably because a large portion of the house cost is in transportation and site preparation (basements, streets, sewer lines, electrical lines, water lines, and landscaping). What if the basements, sewer lines, and all were pre-manufactured so that we could deliver the house with basement, landscaped front and back yards, paved street, and utility hook ups all in one move! What else? How about having the capability of dumping four Olympic sized swimming pools of water on a forest fire or carrying twenty super large mining dump trucks' worth of earth at a fraction of the cost.

These zeppelins could also serve in the area of disaster relief. For instance, we could help evacuate cities like New Orleans after a hurricane and flood. We could transport water to drought-stricken areas or food and housing relief in case of an emergency. In addition, they could deliver pre-manufactured concrete aqueducts or pipe systems (including supports) to places that need water. How about helping with worldwide recycling of our non-renewable resources by continuously traveling in one direction around the globe—possibly with the aid of the jet stream?

Finally, these behemoths could transport liquefied hydrogen fuel to locations within the United States and around the world. and large numbers of mass-produced zeppelins would be required. Transporting liquefied hydrogen via my proposed large-scale zeppelins is by far the least expensive method. They could prove to be 90 percent less expensive than long distance pipelines!

To transport the hydrogen equivalent of all projected gasoline fuel currently used by the United States, we will need about one thousand zeppelins. As you will see later, a 100 percent hydrogen fueled future world will require thousands more.

If we were to construct a large-scale hydrogen producing energy park like the one proposed in the last chapter, the zeppelin would solve the transportation issue. Moreover, the idea of wind- and solar-produced electric energy becomes more viable since hydrogen production and storage is not dependent upon nature. Do you see where this is going? As a result of developing a hydrogen infrastructure, spinoff technologies like the zeppelin could solve some global transportation problems.

Whenever I discuss hydrogen fuel as a replacement for gasoline, it seems like the Hindenburg disaster is almost always mentioned. So, what is it about the Hindenburg that causes people so much concern? Haven't you seen gasoline-powered cars and airplanes explode? If so, why do you think the Hindenburg was so much more of a problem? Is it because the radio commentator said it was the worst thing he has ever seen? Or is it because the Hindenburg film has been played so many times on television that everyone has etched in their minds that hydrogen must be a horrible thing?

The fact of the matter is the Hindenburg is an example of one of the safety aspects of hydrogen fuel. What the Hindenburg showed was that hydrogen is so light that it disperses and floats skyward very quickly. Therefore, when hydrogen burns, it quickly disperses and gives off negligible radiated heat. It's this dispersion property of hydrogen that allowed sixty-two of the ninety-seven people on board the Hindenburg to survive. Of the remaining thirty-five people, thirty-three died from falling or jumping, while only two people died from burns.

Contrary to what you may have heard, the Hindenburg fire was most likely not started by the ignition of leaking hydrogen gas—as implied by the movie *The Hindenburg: The Untold Story*. Evidence suggests that it started by a static electricity buildup and the painted covering, which was highly flammable. Proof of this was demonstrated sixty years after the accident using debris from the crash. The chemistry of the paint formula resembled the fuel for a modern-day booster rocket.

You may remember Ben Franklin's famous kite flying experiment where lightning struck an electric charge collector on a kite that transmitted electricity to the ground. In the case of the Hindenburg, it was an 800-foot-long electric charge collector. When the

Hindenburg became grounded by dropping its landing ropes (which were wet from rain), the experiment was complete, and the electrical charge in the Hindenburg's skin started the fire. (The Hindenburg would have burned and crashed if it had been filled with helium.)

As eyewitnesses have noted, the hydrogen fire took off after the Hindenburg's surface skin started to burn, and the fire was over in less than a minute. The diesel fuel and other heavier-than-air components of the Hindenburg continued burning for hours on the ground.

Perhaps some people are unaware that the Hindenburg had crossed the Atlantic twenty-one times prior to the accident, and its successor, the Graf Zeppelin, a smaller hydrogen-filled airship, flew 144 flights nonstop to and from Berlin, Rio de Janeiro, and New York. It made 650 flights, delivering more than 18,000 passengers safely during the nine years that it flew. It logged over one million miles.

Although a hydrogen leak and static electricity cannot be ruled out as a possible cause of the Hindenburg's fate, there is reason to believe otherwise. The following excerpts are from a May 28, 1998, report that described the results of work conducted by William D. Van Vorst, professor emeritus of chemical engineering at UCLA and Addison Bain, former manager of hydrogen programs at NASA's Kennedy Space Center:

"By studying newsreel footage, examining the chemical composition of the skin and delving into the records of the German firm which built the Hindenburg, they pulled together three independent pieces of compelling evidence indicating hydrogen could not have been the culprit.

> "Newsreel footage contradicts the hydrogen theory," said Van Vorst, who studied individual frames of the footage. "The picture indicates a downward burning. Hydrogen would burn only upward!"
>
> In addition, Van Vorst pointed out, "hydrogen burns with a colorless flame," yet witnesses compared flames aboard the Hindenburg "with a fireworks display." "There is also a remarkable similarity between what can be seen in the newsreel footage of the Hindenburg and the photos and witness descriptions of fires aboard airships containing no hydrogen, but covered with similar materials," Van Vorst said.

Furthermore, the substance used to coat the cotton skin — a process known as "doping" which makes the fabric taut and more durable — was extremely flammable. A combination of iron oxide, cellulose acetate and aluminum powder, "the total mixture might well serve as a respectable rocket propellant," Van Vorst said.

Additionally, the manner in which the skin was attached to the airframe allowed a large electrostatic charge to build up on its surface. When it finally discharged, it did so with disastrous results.

"As a result of the electrostatic activity, the skin became highly charged, and finally passed the current through the skin to the frame. In the process, the skin and its highly energetic doping constituents were ignited, setting off the conflagration," Van Vorst said in the paper.

Analysis of the sister airship, the Graf Zeppelin, being constructed at the time of the Hindenburg accident, indicates remedial measures were quickly undertaken, he said. Calcium sulfumate, a chemical widely used in the textile industry as a fireproofing agent, was added to the doping mixture. The doping compound was further modified by substituting bronze for aluminum. Though heavier, the bronze was far less combustible. Furthermore, the bronze was highly conductive, allowing any static charges to be bled off, rather than build up. Also, the cord holding the fabric in place was impregnated with graphite to make it conductive, thus reducing the electrical potential between skin and structure.

"Clearly, there must have been strong suspicion that the fabric was the real culprit," Van Vorst said. With these changes in place, the Graf Zeppelin, which was also kept aloft by hydrogen, went on to fly more than a million miles without incident.

Furthermore, according to a letter from electrical engineer Otto Beyersdorff, hired by the Zeppelin Company as an independent investigator: "The actual cause of the fire was the extremely easy flammability of the covering material brought about by the discharges of an electrostatic nature." Beyersdorff went on to say he had tested samples of the material in the laboratory "matching the conditions of the accident, which proved the material to be easy to inflame."

Even though letters from an independent investigator and the decision to make design changes in the Graf Zeppelin indicate the company knew the true cause of the blaze, Hugo Eckener, chairman of the Zeppelin Company publicly blamed hydrogen for the disaster. Van Vorst speculated this might have been done to show the United States in an unfavorable light for being unwilling to supply helium for use as the buoyant force. He also suggests it might have been an effort to cover up what proved to be poor design decisions in the choice of doping materials.

Van Vorst, who has spent much of his academic career exploring the use of hydrogen as an alternative fuel, sees the findings as further proof of his contention that hydrogen is a safe alternative to gasoline.

"The public must be made aware that hydrogen may be used as a fuel with the same degree of safety as gasoline," he said.

"With proper handling," Van Vorst said, "hydrogen is no more hazardous than gasoline and may, indeed, be less so."

Although the above UCLA report didn't provide absolute proof regarding the cause of the Hindenburg accident, the possibility of a hydrogen leak or sabotage cannot be ruled out. Current technology could be employed to avert a Hindenburg type of disaster. As mentioned in the report, German engineers did take remedial action for the Graf Zeppelin. In addition to these actions, hydrogen leak detectors are currently available, and much more is known today about static electricity build up and how it can be handled. Therefore, additional countermeasures could be taken if a hydrogen leak were detected. Besides —my proposed zeppelin is an unmanned remote-controlled vehicle that's primarily used to transport liquefied hydrogen. If passenger transportation becomes a spinoff use, helium could be employed.

Some people might roll their eyes and say—yeah sure. This thing is so big it would be impossible to build or mass produce—especially as many as 300 zeppelins per year. Well, history is filled with people who said it couldn't be done. If we had listened to these people we wouldn't have gone to the Moon, built jumbo jet airliners, or the Golden Gate Bridge. To alleviate this concern let me put some things into perspective by mentioning some related facts:

1. The Golden Gate Bridge's towers are 770 feet tall, and the middle span is 4,200 feet long.

2. The NASA space shuttle crawler is 131 feet long x 113 feet wide and is capable of transporting 12 million pounds.

3. The University of Phoenix Stadium has a 403-foot-long x 234-foot-wide 18.9-million-pound moveable field.

4. The University of Phoenix Stadium has a 1/16-inch-thick Teflon-coated fiberglass retractable roof that is supported by two 700-foot-long x 87-foot-high (at its mid-point) trusses.

5. The USS *Ronald Reagan* aircraft carrier is 1,092 feet long x 34 feet wide and weighs 190 million pounds when loaded.

6. The United States learned how to mass produce 2,751 Liberty Ships during WWII, at a cost of under $2 million each (about $20 million in today's dollars). These ships were 441 feet long x 56 feet wide. Each ship was comprised of 250,000 parts prefabricated throughout the country. The average ship took 70 days to complete with the last ship—the SS *Robert E. Perry*—being built in four and a half days.

Let's begin by building smaller zeppelins[1] that can be used to construct my proposed remotely located renewable energy park, thus minimizing the cost of roadways and aiding in the construction process. If we build a zeppelin that's 200 feet in diameter and 75 feet tall, it can lift almost 50 thousand pounds at ground level—the equivalent of lifting 10 elephants weighing 2.5 tons each. Double or triple stacking could also be used for very heavy lifts.

---

1   For those readers who would like to see an illustration of my proposed torus shaped zeppelin please obtain a copy of my book *Reaching America's Destiny*.

CHAPTER 7

SAVING THE WORLD

In the introduction to this book, I mentioned that the hydrogen revolution could act as a catalyst to help poor and destitute people around the world and possibly help to make the world a safer place with help from a worldwide implementation of an advanced form of blockchain technology.

Today's world population is about 8 billion. About 70 percent of that population is living in poverty, and there's currently not much chance of changing their circumstance. With this thought in mind, it's easy to understand how oppressive totalitarian governments continue to survive—and thrive—by keeping their populations under control. After all, with a small percentage of government leaders in totalitarian countries enjoying wealth and modern-day conveniences, it's easy for them to oppress and control the multitudes by keeping them in poverty. Why? Because the multitudes have no money, and it takes money to get power to change the way it is. Can a hydrogen revolution change this situation? If so, will the result lead to a safer world? I think yes, because I believe when people are lifted out of poverty, they will be emboldened to act and change the way their governments behave. With a taste of freedom that financial security brings, they will want more freedom and will not tolerate a dictatorial government. While some might say this is "pie in the sky" thinking, consider how this future might evolve. Perhaps when the poverty situation is solved, a future generation of government leaders might emerge that listens to its people.

With the above being said, let's begin by establishing a basic understanding of economics and how productivity can improve people's lives. I call it economics 101.

Years ago, famed economist Milton Friedman visited China and observed a hundred workers building a road with shovels and pickaxes. When he asked the man in charge why they weren't using modern day equipment, the supervisor said that by using shovels more workers could be employed. Friedman responded by asking, why not use spoons and employ even more workers? If you believe that spoons or shovels are a good way to keep people employed and eliminate poverty, think again.

If 100 workers earned five dollars per day with shovels, then five workers could earn $100 per day if they used modern equipment. Moreover, these more productive workers could probably build the road ten times faster. What this means is that the cost to build the final road would be much less than the original budget, and the five more productive workers could be paid as much as $300 per day or about $75,000 per year. But what about the other ninety-five workers? Won't they be unemployed? Well, the five productive workers could employ them by having them build five houses and provide other goods and services. Since these highly paid productive workers now have enough money to pay back an amortized loan, their money can be leveraged. If the housebuilders use modern tools and equipment, they can also earn a good wage since the faster they build the house the more money they can make. Since the house builders will earn more money than they did when using shovels to build the road, they can buy more goods and services that require more people to be employed. Then these newly employed people also buy houses, goods, and services that eventually allow the ninety-five workers to continue working at a higher wage.

Let me explain what I have just said in terms of the original $500 per day that was used to build a two-mile long road. Let's assume that whoever is paying for the road has enough money for 300 days of road construction per year. This would mean that $150,000 is available per year to build two miles of road. Because of building the road 10 times faster with modern equipment, the two miles of roadway would cost 10 times less per mile. Because of the greater efficiency in building the road, the road buyer might negotiate paying three times as much for 10 times as much roadway. If so, the five productive workers could be paid $300 per day rather than $100. This would mean each of these five workers would receive $75,000 per year in wages. If each of the five workers then bought a $150,000 house, using mortgaged money, $750,000 would be available for construction of five houses. If the houses were built in six months using modern equipment, the ninety-five previously unemployed workers could potentially each earn $7,900 ($750,000/95 = $7,900) per year. If additional houses were constructed for the remainder of the year, each housebuilder would earn $15,800—a far cry from the $1,500 per year they were earning as a shovel worker. When this process is done on a continuing basis, the economy grows, and more people earn more money to buy houses, goods, and services.

Although the above example is simplistic because it ignores such things as equipment and materials cost, it makes the point about productivity and leveraging money. By being able to produce goods and services faster and cheaper using modern equipment and automation, we can increase employment rather than decrease it. When banks lend money, they allow more money to be spent. As a result, the money is leveraged to a higher short-term value that stimulates economic growth. If the 100 workers in the road building example continued to earn only five dollars per day—or about $1,500 per year—they would earn only enough money to buy food to survive. The idea of having them mortgage a house would be out of the question. But this is the circumstance that most people in the world find themselves today. We can do much better, and here is where the hydrogen revolution comes into play.

Do you remember what I said about farmers and gas station owners making money by generating solar-derived electricity and selling electrolysis-derived hydrogen from water? Well, how about applying this idea on a worldwide basis, where impoverished people have access to land, sun, and water.

Jack Welch was the former CEO of General Electric and arguably the greatest private business CEO of all time. When Jack was asked during a televised interview if China was an economic threat to the United States, he said no, since from his viewpoint China represented more than a billion new consumers. As China's economy grows, so will the number of consumers who can buy goods made in the United States. In other words, economic growth in the underdeveloped world can economically benefit the United States.

Do you see where I'm going with this? As I see it, worldwide economic growth could have a positive economic effect here in the United States, while at the same time the United States could have a positive influence on the rest of the world. Perhaps the most important result could be world peace and—please allow me to say at this point—Christianity. Since the United States and the rest of the world will benefit, why not redirect our foreign aid money to promote this effort? In fact, we could use the Peace Corps. However, I recommend the US military become involved. After all, the purpose of having a military is to not only fight a war, but to prevent one from happening. Preventing a war from happening is generally thought to be the result of military strength. However, with the possibility of weaponized viruses and cyber-attacks on infrastructure, the idea of conventional war is changing dramatically. By using the military for peaceful missions, we might be able to directly change enemy hearts and minds. With that thought in mind, let's not measure our military strength in terms of numbers of aircraft carriers, howitzers, fighter aircraft, and tanks, but rather in terms of evolving technology and direct contact with potential enemies. After all, an enemy satellite or drone could track and destroy an aircraft carrier and guided missiles can knock out fighter aircraft or tanks. In that sense the military might need to be reimagined. While many countries might not accept the US military performing peaceful missions in their country, if they see the success a hydrogen revolution is having on other countries, it might get our foot in the door. If the hydrogen revolution works in non-threatening countries, perhaps threatening countries might accept our offer of help—especially if their populations start migrating to other countries. Keep in mind that by raising the standard of living of poverty-stricken people, they might in turn purchase goods produced in the United States. Hopefully they might also influence their leaders to change their ways.

In case you're asking why a military-style peace corps rather than the civilian Peace Corps, I have six reasons. The first reason is funding. By restructuring the military away from thinking about obsolete methods of warfare, money will be freed up to go toward helping other countries economically. The second reason is security. The military is better equipped to protect the participants—especially in hostile countries. The third reason is to train our troops to operate in a variety of environments. A fourth reason is to put our military prowess and technology on display for the world to see—like, for instance, satellite surveillance and microwave energy from space. Believe me when I say our military presence is awesome. I know this from my first-hand experience in Vietnam. The fifth reason is quick mobilization. The US military has the ability to immediately set up tents and other facilities at a moment's notice. This is a very impressive thing to observe. The sixth, and in my opinion the most important reason, is to spread Christianity. Since I'm assuming that our military troops will be required to act in a way that reflects well on our country, there will probably be some Christians members of the corps who will see an opportunity to spread the gospel.

How does a military-style peace corps actually raise the standard of living of poor and destitute people? To begin, the peace corps groups will have access to lots of money—about $200 billion per year from a more than $700 billion military budget. This money could be used not only to support the peace corps personnel but to also purchase American-made solar panels, electrolyzers, hydrogen dispensing equipment, and other things that I'll mention later.

CHAPTER 8

RECYCLING AND RENEWAL

I magine a world where today's population of about 8 billion doubles to 16 billion, and everyone is consuming non-renewable resources and disposing them in landfills at the end of their useful life. To make matters worse, consider the fact that 70 percent of today's 8 billion are poor and destitute and don't consume non-renewables at anywhere near the rate consumed by the other 30 percent. To give the reader a sense of timing, today's world population of about 8 billion people is predicted to exponentially grow to about almost 10 billion by the year 2050, and 16 billion by the year 2085. Most of this exponential population growth will consist of poor and destitute people and is well within the lifetime of your children and grandchildren. With these numbers in mind, perhaps my idea of raising the standard of living of the world's poor and destitute population might be considered a bad idea. On the other hand, this predicted population growth assumes that nothing is done to slow it down. With my plan, the population growth will be voluntarily slowed to about 10 billion rather than 16 billion in sixty years. Why? Because the people in prosperous industrialized countries tend to have fewer children. In some countries like Japan there is a negative population growth. So, by raising the standard of living of billions of the poor and destitute a natural slowdown should occur. Therefore, from this point forward I'm going to assume a world population of about 10 billion by the year 2085.

To put this into perspective, consider these facts:

1. Half of the world's almost 8 billion people live on less than the equivalent of two US dollars per day.
2. Farmers are abandoning about 27,000 square miles of farmland each year because of a lack of water and soil degradation.
3. The world's fresh groundwater is being consumed about ten times faster than it is being replenished.

4. Many underdeveloped, non-industrialized, countries are slowly becoming industrialized and are consuming more and more of the world's non-renewable resources.

If there are solutions to these problems, we need to start implementing them now, since time is not on our side.

As you are well aware by now, my proposed solutions for world peace and prosperity begins with hydrogen being a catalyst or launching pad. But as previously mentioned other factors need to be taken into consideration like depletion of non-renewable resources and the availability of food, and water. While it's possible that by the year 2085 we will have begun mining the Moon and asteroids for non-renewable materials we will need an interim solution that I will call worldwide recycling and renewal.[1] With regard to food, I recommend hydroponics, which will be described in the next

---

1   After reading John Lewis' book *Mining the Sky*, I found that the following amount of non-renewable materials available in our asteroids is estimated to be about: 1,650,000,000,000,000,000 thousand pounds of iron; 115,500,000,000,000,000 thousand pounds of nickel; 8,250,000,000,000,000 thousand pounds of cobalt; 24,750,000,000,000 thousand pounds of platinum; Plus, large quantities of other materials such as gold, silver, copper, manganese, titanium, rare earths, silicon, and uranium.

chapter. With groundwater aquifers being depleted ten times faster than they are being replenished, let's again be creative. How about either importing fresh water from my proposed OTEC rigs or build grand canals and pipelines to divert fresh water from water rich regions of the Earth to water poor regions of the Earth.

Since my assumed population of 10 billion probably won't occur until the year 2085, I need to point out that nuclear fusion energy (not to be confused with nuclear fission energy) might be cheap and abundant at that time. In which case, what I'm going to propose for this sixty-plus-year timeframe would probably need some modification if that were to happen. Moreover, taking what I've just said into consideration, I would be remis if I didn't forecast a solution beyond 2085. As I see it, Earth's population, at that time, will most likely begin moving to huge space satellites surrounding the Earth and/or other planets in our solar system. Beyond that, Earth's population may eventually seek refuge in another solar system I say this because besides mining asteroids, there is enough nuclear fusion fuel (called helium 3) in the upper atmospheres of the gaseous planets to provide enough energy to sustain a population of 30 trillion (that's not a typo). Moreover, the asteroid Ceres may contain five times as much fresh water as our planet Earth. So, not to be shortsighted in my hydrogen-related proposals, I describe my vision of what this distant future might be like in the epilogue. In fact, the hydrogen-based proposals, described in this book, might set the stage for this future possibility. It's called sustainability, which can be described as the ability of a growing population to provide enough materials, food, and water in a recycled way with minimal or no waste. However, there needs to be an almost infinite supply of energy for this to be possible. In this regard solar, wind, and nuclear *fission*–derived energy might fit the bill for the next sixty years. However, helium 3–fueled nuclear *fusion* energy will most likely be the ultimate, and almost unlimited, source of future energy.

Keeping what I've just described in mind, let's now address the near term need to preserve non-renewable materials as it relates to helping poor and destitute people. If all the world's population consumed oil at the rate currently consumed by the United States, we would deplete Saudi's super giant oil reservoirs in months rather than tens of years. If so, then oil and other fossil fuels are not the answer to a growing world population that enjoys the material and health care benefits of our modern industrial age. That's why I am such a proponent of using hydrogen fuel derived from nuclear and renewable energy and intelligently utilizing the earth's vast supply of water. As previously mentioned, although current water-cooled nuclear fission reactors could provide a significant amount of electricity, we need to have breeder reactors and/or fusion nuclear to provide for a future that consumes a gargantuan amount of power. Regarding a depletion of non-renewable materials, how about implementing worldwide recycling, using globe-circling zeppelins.

While making low-cost basic goods that poor people can afford seems like a good idea, perhaps we should supplement this idea by providing recycling service jobs. At some point worldwide recycling of non-renewable material resources may not be an option. As I see it recycling what we already have may become an economic necessity. Repeating what I've been saying to this point, if a future population of 10 billion people or more were to use non-renewable resources the way that the industrialized world uses them today, most of these resources will quickly become depleted and end up in landfills.

From an idealistic engineering perspective, worldwide recycling needs to begin with new product design. Currently, many design engineers think more about making a new product that will sell, than what will happen to the product when it wears out. In fact, many times there is little concern for the repair mechanic that needs to service the product. With recent advancements in robotics, it would be prudent to design new products with robotic assembly in mind. By doing this the designer could also consider robotic disassembly. In a perfect recycling world, products would be produced on an assembly line, with worn out products being returned to the factory for recovery on a disassembly line.

Since the original product manufacturer knows more about the pedigree of the original product's material and assembly process, that manufacturer would be best able to perform the disassembly and reuse process since they would have access to all the original parts. These part and subassembly manufacturers could then return products to the originating companies to the point of raw materials processing.

While you might again be saying this is "pie in the sky" thinking, let's examine the advantages to companies that do participate in this process. First, products assembled and disassembled by multi-purpose robotics tend to be less expensive—even when compared to low-cost manual labor. Second, if the manufacturer planned for disassembly of obsolete products, with the potential for reusing the obsolete product's materials and subassemblies in their new products, they could further reduce costs. Third, the product manufacturers would use more non-corroding materials such as structural plastics, carbon composites, aluminum, stainless steel, titanium, copper, and brass, which would tend to make their products lighter weight, longer lasting, and potentially higher quality. Last, but not least, the product manufacturers could restore (remanufacture) obsolete products for low-cost resale to emerging countries. In fact, one scenario may be to make new products from old parts. Even though selling remanufactured automotive parts is a big business today, what I am talking about is remanufacturing an entire automobile (or other similar products). People in emerging countries are not as enamored by up-to-date products as they are by being able to afford a product that they previously couldn't afford.

You may now be asking yourself how are we going to rescue the obsolete products and return them back to each manufacturer in the world without incurring a high cost and complexity? The short answer is by using the FedEx and UPS systems. These companies generally deliver packages to central locations for sorting, and then ship the sorted packages to their destinations. Some packages that are destined to go to a nearby location are actually sent to distant central locations first. Believe it or not this turns out to be the most efficient way to do it.

But doesn't transporting all these products cost a lot of money and use a lot of fuel? Yes, if done by conventional means like air cargo planes and trucks. What's my answer? That's right, you guessed it. Let's use zeppelins.

So, what does all this have to do with the billions of poverty-stricken people around the world who are trying to survive on the abovementioned degraded farmland? The farmers could be employed to disassemble and clean products for shipment back to original manufacturers. They would be the Federal Express central processing and sorting locations.

Let's just begin by looking at materials that are currently destined for landfills. With the hydroponic facilities, that I'll describe in the next chapter, enclosed spaces and electricity would be available for this purpose. While current recycling generally includes high-valued materials, like aluminum engine blocks, my proposal goes further. For example, how about specializing in preserving automobile chassis and selling them to other specialists who restore automobile axels and wheels. By using the internet almost all non-rusted materials can be preserved, identified, and moved from one location to another via zeppelins. Eventually a large percentage of materials can be refurbished and sold for a profit. Better yet, some innovative people might use these restored parts to make new products. In effect poor and destitute people could make a decent living selling recycled materials, hydroponically derived food, and—let's not forget—hydrogen.

CHAPTER 9

HYDROPONICS

In addition to worldwide recycling to preserve non-renewable materials, let's look at a potential food shortage problem. With depleted farmland, and many more mouths to feed in the future, how about growing food hydroponically?

What is hydroponics you may ask? Hydroponics is a form of gardening that instead of soil uses a solution of water and nutrients. Plants – even trees—tend to grow faster, and if the system is enclosed, the process can be implemented year-round.

Let's compare productive farmland and hydroponics. The gross income from one acre of highly productive farmland in California's Coachella Valley is about $8,000 per acre per year. In comparison my proposed two-level hydroponic farm building can generate a net income of about $110,000 per year in in the same one acre space. In addition, 70 to 95 percent less water would be consumed in the hydroponic system, and no pesticides would be needed. Moreover, liquid fertilizer could be piped from sewage treatment plants or fishponds at little or no cost. For example, a hydroponic farm can grow five crops of corn per year with a yield of about 14 ears per square foot. A 300 square foot space would yield 4,200 ears of corn, or what could be enough to feed one person for about one year.

So, what does this two-level hydroponic farm building look like? Well, the top level is a glass-enclosed hydroponic farm, while the lower level can serve many purposes. One purpose is to geminate plants and grow additional hydroponic food using artificial light. It could also serve as condominium style housing and/or manufacturing facilities. As mentioned in the last chapter, perhaps some of the available space could be for recycling facilities.

My estimate is that my proposed two-story concrete and glass building 1,200 feet long and 110 feet wide would cost about $12 million. As part of the proposed building, vertically placed sun-tracking solar panels could generate hydrogen fuel that could be sold to a world market. My calculations show a net profit of $350,000 per year after subtracting the energy used by about 120 building inhabitants. An additional profit can be realized by selling food. Using sweet corn at 14 ears per square foot per year, a two-acre (i.e. 2 actual crop acres per hydroponic building) hydroponic crop could produce 1.22 million ears. If fresh sweet corn sells for about $0.50 per ear at a grocery store or farmers' market and $0.30 per ear wholesale, the resulting net income (after feeding the building's inhabitants) would be about $250,000 per year.

Hydroponically grown crops—including hydroponically grown fruit and nut trees—could be the wave of the future. With much improved crop yields per land use and potentially reduced investment cost, it only makes sense. Water usage would be greatly reduced, and weeding would be unnecessary. Pesticides would be eliminated, and transportation and farming equipment, with its associated fuel cost, could be minimized. Instead of nearby towns and cities paying to dump gray water from sewage treatment plants into nearby rivers and streams, you could use it to fertilize the hydroponic crops and possibly pay the hydroponic farmers for it.

To go even further, I propose arranging the hydroponic buildings in a way that forms a nucleus that serves a greater purpose. That purpose being to improve the lives of poor and destitute people around the world. Let's assume two-mile by two-mile squares arranged in a checkerboard fashion and sixty south facing hydro-ponic buildings placed at the center. At the corners of the checkerboard squares we place sewage and waste disposal facilities, hydrogen production, and fuel cell electric power generation, a reservoir and water supply system, and an area to make concrete and grind or melt scrap materials. In part of the center portion of each checkerboard square, we could have a zeppelin transpor-tation system, a runway for small aircraft, and a village center. The village center could include schools, church-es, a medical facility, a mini-Wal-Mart, McDonalds, rec-reation center, movie theater, auditorium, grocery store, bank, office building, and more. Although a checker-board pattern allows for monotonous replication and expansion, many more attractive versions—including forests and lake areas, for example—might be possible.[1]

Do you see where I'm going with this? I'm try-ing to find a solution which will allow the poor and destitute of the world to become prosperous and live their lives in a fashion similar to how most people live in the United States. The hydroponic town complex described above forms the basis for my concept. I'll be more specific in the next chapter.

---

1   For readers who are interested in seeing illustrations of what I'm proposing please read my book *Reaching America's Destiny.*

CHAPTER 10

TEN BILLION PEOPLE

As previously mentioned, I'm assuming a world population of about 10 billion by the year 2085. During that sixty-plus-year timeframe, I recommended in chapter 7, a US-sponsored military-style peace corps to help raise the standard of living of poor and destitute people around the world. However, for practical purposes the peace corps personal will concentrate on raising the standard of living of only a small fraction of the people who need help. The thought being that there will be a self-sustaining spillover effect that could help billions more people over time.

In the scenario I'm going to describe, I'm assuming that the peace corps will provide what might be called stimulus money to get the ball rolling so to speak. To be specific I'll assume the peace corps focuses on five hydroponically based building complexes each year and provides a total of $200 billion each year to the process. Since the current US military budget is in excess of $700 billion, this would be money well spent for the security it might provide. Besides, much of the building materials, that would be required, would be made in the US.

So, let's begin an excursion into my imagined—and idealistic —future where about five previously described two-mile by two-mile hydroponic building complexes are constructed each year. My concept specifies sixty hydroponic buildings per complex. These sixty-unit complexes would be located near the center of an eighty-mile by eighty-mile land area (6,400 square miles, or what would result in 12 percent of the world's farmland at the end of a sixty-year timeframe).

As mentioned in the last chapter, the estimated cost for each of these 264,000-square-foot hydroponic buildings is about $12 million—including vertically attached sun-tracking 26 percent efficient solar panels. Also, as previously mentioned, each building generates a net profit of about $600,000 per year by selling hydrogen fuel and food. This would be a net profit of about $36 million per year per sixty-unit hydroponic building complex.

With that being said, let's make some other assumptions. The first being that five sixty-unit hydroponic building – starter—complexes are built each year and— as previously mentioned—they are fully paid with peace corps funding, A simple calculation at this point would show that each of these five hydroponic building complexes might have a starting total population of about 36,000 (7.2 thousand people per sixty-unit complex). But as also previously mentioned, each complex is surrounded by 6,400 square miles of land that could be used for self-sustained growth. If three additional hydroponically based complexes are added at each of the five startup locations each year for sixty years, the result would be a prosperous population of almost 2 billion people by the sixtieth year and more than 4 billion by the 120th year. However, in my analysis, the population in the area surrounding each 7,200 -person sixty-unit complex would most likely grow as each year passes (Re: per my previously discussed economics 101 using leveraged money) to about 60,000.

At this point, you might be asking how can one initially constructed startup sixty-unit hydroponic building complex eventually self-expand/spillover into three more each year? Well, let's begin with each of the five startup complexes having $40 billion provided by the peace corps. If $39 billion of that money was spent building solar/hydrogen production facilities, a gross income of about $3.3 billion per year could be realized. My estimate is that the materials cost alone for each sixty-unit complex is about $720 million. In which case we use the solar hydrogen income to build three additional complexes each year at a total expenditure of $2.16 billion. This leaves about $1.14 billion per year to pay for construction labor and other growth expenditures like expanding each complex to accommodate more people. Plus, as you may recall, each sixty-unit hydroponic complex adds another $36 million per year by selling hydrogen fuel and food.

For those readers who want to see specifically how I arrived at my numbers, I suggest making your own estimates and assumptions with an Excel spreadsheet. Of course, if an influential government person or billionaire entrepreneur was really interested in making my plan a reality, I would be more than willing to provide year-by-year details. These details would include an economic payback and illustrations that show how a government, or privately sponsored, prototype demonstration can get the plan started. Think of Native American and/or government owned land in the western part of the United States.

Adding to what I've just proposed, it's my opinion that the vast majority of poor and destitute people, who will occupy these proposed facilities, will come from other countries. As I see it, most will probably come from oppressed countries like China, Russia, and North Korea in an attempt to gain freedom and prosperity.

As an illustration, let's assume one possibility. That possibility is that many people will migrate to Australia. It has less than 26 million people and vast amounts of unused land that could accommodate billions more. Australia's population has remained low because of its harsh hot and dry inland conditions. The Australian's call it the outback for obvious reasons. Having the least amount of rainfall of any country in the world, except Antarctica, Australia needs fresh water to make the outback habitable. After studying many ideas—like grand canals—I have an idea that might not only work but might also pay for itself. Remember the OTEC discussion from a previous chapter? Well, how about using OTEC systems to create hydrogen and fresh water? After doing the math, I found that by strategically placing one hundred OTEC nuclear systems off Australia's coasts at a cost estimate of $6,000 per kilowatt for a 100-megawatt system (an estimate based upon previous studies), we will spend about $600 million. If we add five 20-megawatt modular nuclear power systems at a cost of $4,000 per kilowatt to generate surface water heat and electricity, the total cost per system might be about $1.0 billion. If we add an average of $1.0 billion to each OTEC system to route freshwater pipelines into strategic locations of Australia's interior, the result will be almost 5 trillion gallons of fresh water per year. If we assume Australia's population increases to 2 billion people who make and sell hydrogen fuel, hydroponically grown food, and recycled products, 5 trillion gallons will work. Besides, the income from selling OTEC-produced hydrogen at $8 per kilogram to passing ships will pay for the entire project. If we assume an amortized loan of thirty years at 3 percent, each system will have a positive cashflow of $10 billion per year after the amortized loan expense—or about a 5.1 percent return on investment. But wait, how about selling 5 trillion gallons of fresh water per year to 2 billion people. At a US average cost of $0.025 per gallon of fresh water, we can add another $125 billion per year on top of the $10 billion.[1]

I know you may be rolling their eyes in disbelief. Okay, I get it. But just remember: *"Some people see things the way they are and say why. I dream of things that never were and say why not?"*

_____

1   As a matter of interest, it takes about 2.5 gallons (9.43 kg) of water to produce 1 kilogram of hydrogen. Thus, about half of the 5 trillion gallons of fresh water per year will be needed for this purpose alone. Pipelines make more economic sense than zeppelins to transport water, and other heavy liquids.

CHAPTER 11

PEACE AND PROSPERITY

Putting all of what I'm saying together in a form that results in world peace and prosperity consider this. Imagine the new idealistic, self-sustaining hydroponically based towns and cities, that I've described in the last chapter, being a part of almost every country in the world—including the United States. Now, imagine that an advanced style of blockchain technology has become the world operating system for voting, currency exchange, and personal affairs. Could these changes result in world peace and prosperity? Well let's examine how. But keep in mind, my imagineered world needs to be a slow and gradual multi-year process where the world's population will most likely grow from about 8 billion to about 10 billion (in about sixty years).

The world I'm proposing might cause people living in countries with oppressive governments, to renounce their citizenship and move to another country. The extent to which this might happen is anybody's guess, but I wouldn't be surprised if it didn't happen on a massive scale. For example, if we modified Australia's outback, per my proposed OTEC idea, two billion people could seek refuge there. Even in water-rich free countries with open land spaces, like Africa, South Korea, South and Central America, Canada, Mexico, and the United States, migration opportunities might be compelling. However, since water is generally not evenly distributed in most of the countries mentioned, grand canals and/or water pipelines might be required for water redistribution.

For the United States in particular, we have a significant amount of Native American and desolate, federally owned land. We could build a grand canal[8] and water pipelines to supply needed water to these lands and help our own Native American people become prosperous. We would be able to build prosperous new creatively designed cities and towns that could provide an outlet for our many currently overcrowded cities. These towns could also be a solution to illegal immigrants who have and are currently crossing our southern border.

Of course, if this migration were to take place, it would need to be with the consent of the countries involved. After all, the idea of billions of migrants with varied languages and beliefs might pose some difficult problems for any country. As I see it, each country affected would most likely have stringent entry conditions—including the numbers that would be accepted. For instance, admission into the United States would most likely be contingent upon requirements like the rate and numbers permitted per year, learning the English language, taking a course in US history, pledging allegiance, and agreeing to a multi-year contingent path to citizenship. These requirements could be helped as part of my proposed military-style peace corps directives. However, because these migrants would be self-supporting, they would not create a welfare taxpayer burden. In fact, they would significantly add to an existing tax income.

As previously mentioned, I believe most migrants coming to the United States would likely be coming from oppressive countries like China, North Korea, Iran, and Russia. If this was the case, most of them would be well aware of what a lack of freedom means and would not tolerate another oppressive and controlling government. After arrival they would want to learn the English language out of necessity. Since the idea of non-English speaking cities and towns would be a detriment to the functioning of the United States, a long-term gradual and orderly buildup would be a necessary part of the process. For example, the children of these migrants would be taught in English speaking schools—a process not too different from what earlier migrants that came to America experienced. While some people might not appreciate more migration, just remember the United States is a melting pot of races, languages, and religious beliefs. As such, the acceptance of migrants from oppressed countries might transform the United States and other free countries in a positive way—one of which is world peace.

Since each of my proposed self-sustaining hydroponically based towns and cities produce and sell hydrogen fuel, food, recycled goods, and other commodities as time passes, they could collectively influence government policy. The idea being that each country in the world could theoretically become internally controlled by the majority of its citizens using an advanced blockchain voting system. Believe it or not, even in supposedly free countries like the United States the people have little or no say in government affairs. Why? Because of corruption, fraudulent voting, bias media, paid demonstrators, lack of term limits, and control of the government by special interest groups/lobbyists. Have you ever wondered why elect-ed politicians and appointed government leaders get to be multi-millionaires while in office? Have you ever wondered why politicians break campaign promises? Have you ever wondered why government-appointed leaders of the FBI, CIA, and other agencies have little or no experience and place political agendas ahead of their job responsibilities? In my proposed advanced blockchain system we the people will know who in our government is receiving money (bribes) from special interest groups/lobbyists and who the special interest groups/lobbyists are. Perhaps I should say more about this, but it would require another book. The question is, can my proposals save America from itself becom-ing an oppressive dictatorship?

Let's now look at an oppressive dictatorship gov-ernment. Consider China's apparent desire to take over the sovereign country of Taiwan. If my hydrogen and blockchain ideas were in effect, many of China's cit-izens could ban together and conduct an economic protest. Perhaps an economic protest would never reach this point in the first place if war-minded dicta-tors were voted out of office using a fraudulent-proof blockchain voting system. However, in the case of dictatorships like China, voting their leaders out of office is more complicated than with a republic like the United States. In this regard, let's review China's voting system and how blockchain voting and citizen action could affect change.

China's citizenry currently has no say in who leads their country. Instead, the leader is chosen by committee. This committee is currently comprised of 2981 members who are appointed by the Chinese Communist Party (CCP) to control who would be elected to replace the leader in the event of death or for some other reason. There are about one million villages in China, however, that have duly elected officials that can be recalled by the CCP if they don't abide by the party's directives. Unfortunately, these village voters can only affect actions related to their villages and have little or no influence over the actions taken by their country. What can my proposed hydrogen/blockchain system do to help this situation, you might ask? Well, if half of the million Chinese villages were to become part of my hydrogen fuel, food, and recycled materials supply system, they could collectively have a voice. If these villages became self-sustaining enclaves which were capable of banning together and blockchain voting for advocates of freedom from tyranny, they could effect change. Those who joined the system, would set an example, within China and influence other villages to side with them. A collective protest could result in changing how their government operates. Because millions of these citizens would have control of a portion of the country's hydrogen fuel and food supply pricing, even a dictatorial government most likely would be influenced. Moreover, what if a large number of Chinese people renounced their citizenship and migrated to free countries like the United States? They most likely would leave some of their relatives behind with whom they could communicate with and expose the propaganda from the Chinese government and eventually hold it accountable for blatant lies. Is this just wishful thinking on my part? What do you think?

In today's world, the poor and destitute people in China have very little economic sway, and thus are powerless to influence their leaders. In my proposed world, the poor and destitute people would be collectively self-sustaining with money and influence. To illustrate my point further, assume a quarter billion poor and destitute people in China were able to enjoy a modern-day income of $50,000 per year for a family of four. In addition, imagine they own their own house and produce their own food and energy. In this case, $50,000 per year times a quarter billion families equates to a total income of $3.1 trillion per year. If this income is taxed at an average of 10 percent, they would collectively control $312 billion in tax revenue plus having some control over the price of hydrogen fuel and food. In effect they could go on strike. Moreover, these people could afford satellite transmission of news from outside their country that is not government-sponsored propaganda. As a result, I believe knowledge and truth can be a powerful force that can effect change.

Another interesting point has to do with the amount of money associated with a worldwide use of hydrogen if and when the world population reaches 10 billion—sixty or more years from now. It's a world where 2 billion (or more) less people are no longer poor and destitute. In other words, they enjoy a standard of living that resembles that of the United States. The United States currently has a population of about 330 million people and consumes about 936 gallons of gasoline per person per year. This equates to almost $1 trillion per year at $3 per gallon of gasoline. If 6 billion of the then existing 10 billion people were to consume gasoline (or the equivalent of hydrogen), at this same rate, this expenditure alone would total almost $17 trillion. If 30 percent of this money is collectively controlled by my proposed hydroponically based towns and cities, imagine how they could collectively influence world affairs and demand freedom.

Again, I want to remind the reader of this book that my proposed hydrogen-based world economic model wouldn't be viable for another sixty years. I also see the United States taking a lead role by implementing my proposed military-style peace corps that uses part of its military funding to supplement the cost of equipment, building materials, and labor. With this multi-year plan, the logistics could be modified according to changing times and technologies, but the focus would need to be on keeping the end result in mind. However, as time goes by, I envision almost all of the world's countries participating with their own peace corps and funding. I see this happening if they see the potential for world peace, dramatically reduced poverty, and the sustainability of a growing population.

Finally, rather than trying to achieve a one-world governing dictatorship—as some people are currently proposing—how about a multi-country world in which the citizenry in each country maintains peace and world order? As I see it, hydrogen fuel, blockchain technology, hydroponics, world prosperity, and world peace can be interconnected. It's why I've titled this book **"The Hydrogen Connection."**

EPILOGUE

Now you know how the hydrogen revolution and blockchain technology could bring world peace and prosperity. As an anonymous person once said, "if want to go someplace without a plan—or a road-map—you probably won't get there." Without planning for the eventual depletion of fossil fuels, depletion of non-renewable materials, a lack of food, and our exponentially increasing population, chaos, extreme poverty, and devastating wars might result. In fact, it might be the tribulation phase that the Bible predicts will happen before Christ's return to rule the Earth for a millennium. There is evidence, however, that God has delayed the tribulation phase in the past to allow more people to respond to the gospel—WWII is an example. In my plan, some members of the military-style peace corps will act as quasi missionaries. As a result, I am assuming a further delay of the tribulation as more of the world's population responds to the gospel of Jesus Christ.

With this in mind let's begin an excursion into the future with the idea that in sixty years or more the United States becomes prosperous and more influential in world affairs. In that sixty or more years, the world's population might have grown to about 10 billion, and nuclear fusion energy most likely will have begun to emerge. We might also have developed a Moon and asteroid mining program and begun transporting high-valued materials—including platinum and helium 3—back to Earth. (As previously mentioned, helium 3 is a game changer as it is readily available in space and is the most appropriate fuel for nuclear fusion— not fission—energy.) As I see it, transporting materials from the Moon and asteroids to Earth will require an intermediate step. A robotically built orbiting space station will be needed to process the Moon materials in zero gravity. The space station will most likely be a very large cylinder that rotates to simulate gravity on its inner surface (a NASA proposal many years ago) and may also (for economic reasons) serve a secondary purpose of transmitting solar derived electrical energy to Earth in the form of microwaves.[1]

With these abovementioned conditions as a starting point, let's go further into the future to see how technology advances and human lives are changed. Recent history has shown that many of our modern-day inventions were derived from science fiction books and movies. Although Leonardo da Vinci didn't write a conventional book, many illustrations in his sketchbook show his unusual ability to predict future inventions (weapons of war, flying machines, water systems, and work tools). Likewise, Jules Verne predicted battery-powered submarines, helicopters, and rocket ships. In recent times *Star Trek* gave us some ideas that may still come to pass—like warp speed, phasor weapons, holodecks, molecular transportation, and tractor beams. While these future predictions may serve to capture our imagination, they are more geared toward entertainment.

While my predictions may be less entertaining (I'm not predicting encounters with unusual beings like Klingons or space battles that endanger planet Earth) and my predicted spaceship varies considerably

---

1   Solar energy derived from space is 8 to 10 times more effective than solar energy derived on Earth, and microwave transmission to Earth will have a negligible effect on air traffic.

from the Starship Enterprise, I can assure you that they are equally intriguing. Perhaps they are more entertaining since they are extrapolated from currently known science. To me, this is an important aspect since it's easier for most people to identify with and understand.

I agree that predictions about the future are probably best described in the form of a science fiction story. So, that's what I'll do. My story is excerpted from a book that I may (or may not) publish. It's titled "Epsilon Eridani," and it's about a multi-generational spacecraft that travels to the Epsilon Eridani solar system to investigate an Earth-like planet. I've gone out of my way to make the book exciting and entertaining, but contained within its pages are many coincidences related to Earth's formation, human evolution, religious beliefs, and history; and these coincidences differ considerably from what many people have been led to believe. Be that as it may, the following is excerpted from the beginning of Chapter 1 and another chapter that describes some things I think you will find quite fascinating:

> Hello, my name is Britt Harvath. It's the year 2360. I'm one hundred and thirty-six years old. I live on a very large spacecraft, and I'll be your tour guide. We are currently orbiting

the Epsilon Eridani planet that contains animals and what appears to be intelligent but primitive life.[2] It has taken one hundred ten years for our spacecraft to get here from planet Earth, and my great-grandson Jason Harvath is currently organizing an exploratory team to investigate what's on the planet's surface.

Oh, you want to know more about our spacecraft. Well, as you can see, I'm sitting on a bench in a park located on the top layer of the inside surface. In case you're wondering why I'm not floating around in zero gravity, it's because our cylindrical spacecraft is rotating so that objects—like me—are held onto the inside surface by centrifugal force. It's like the force you feel when you tie a ball onto the end of a string and spin it around.

---

2   After giving much thought about life existing on other planets in the universe, I've concluded that no other life exists on other planets unless God put them there. I discuss my reasoning in great detail in my book *A Different Point of View*.

Before I describe the internal parts of the spacecraft's design, I need to tell you that it's sixteen miles in diameter and forty-eight miles long. Pretty big, huh? It's made almost entirely from materials obtained from the asteroid belt located between Jupiter and Mars. Our oxygen and nitrogen atmosphere were derived from processing materials and from Jupiter's moon Titan. Our water was obtained from the asteroid Ceres and our helium 3 fuel was derived from Jupiter's upper atmosphere. In fact, since we left Earth, more than ten thousand, sixteen-mile by forty-eight-mile satellites—pre-spacecraft—have been robotically built, and Earth's population is migrating to them as soon as they become available.[3] When I tell you about how our spacecraft's living spaces are constructed, you will understand why this migration is happening so rapidly.

---

3   You will find an illustration of my proposed satellite/spacecraft in my book *Reaching America's Destiny*.

From what scientists have determined, we could build as many as 12 million of these huge satellites from the enormous amount of materials available from the asteroids in our solar system. With each self-contained satellite being capable of comfortably and luxuriously housing as many as 2.5 million people our limit could be 30 trillion inhabitants. And now that I see the materials available here in the Epsilon Eridani solar system, it appears that human population growth could be endless.

To understand the internal design of our spacecraft you need to come with me to one of our museums where a high-definition hologram describes our spacecraft better than I can do with words alone. Would you like to go there in an electric vehicle on our roadway system or using our maglev train? –Since you are not in a hurry, I recommend the electric vehicle since it will give you a better view of our upper-level common area. I'll signal one of our electric vehicles to pick us up.

Boy, that didn't take long. Please get in and I'll tell the vehicle where we need to go. The first thing I want you to notice is our blue sky and wispy clouds. It's done with a combination of a specialized form of lighting, a translucent surface, and water vapor. It's more than five miles above our head, so from our vantage point, it looks like it did when I lived on Earth. We simulate night and day on a twenty-four-hour basis, and at least once a week we have a preplanned weather occurrence like a rainstorm. In the mountain zone we simulate snowstorms that keep the ski trails fresh. Notice the fenced in area we're now passing. It's one of our wild animal preserves. Safaris are available on a daily basis, if any of you would like to go on one.

Yes, there are numerous uniquely designed golf courses. In fact, we are going to pass by one in a few minutes. Yes, again, we have lakes, rivers, streams, waterfalls, flower gardens, and forested areas. We also have schools, universities, sports stadiums, hospitals, churches, and office buildings. Okay, we've arrived. I'll send the vehicle back to its underground parking space and retrieve it when we are ready to leave. By the way, that vehicle is powered by a miniature helium 3 fusion electric system. Pretty cool —eh?[4]

Okay, here we are at the spacecraft hologram. As you can see, the thick outer shell of the spacecraft habitat was designed to protect its inhabitants from such things as meteors, solar flares, and cosmic radiation. Obviously, the inside part of the satellite is immune to Earth's problems of hurricanes, earthquakes, tornados, ultraviolet and cosmic radiation, volcanoes, tidal waves, a rising ocean level, and magnetic field reversal. In addition,

---

4   When a helium 3 atom fuses with another helium 3 atom, a positively charged proton is released. If this proton passes through a magnetic field, electricity could be produced directly.

with on-board sensors, our spacecraft is capable of maneuvering away from asteroids, comets, and potentially damaging space debris.

The main thing I want you to notice are the stacked circular structures located beneath the top surface common area. They are stacked twelve high, and there are one hundred and five thousand of them. They are eight hundred feet in diameter and one hundred feet high, and each family living on board owns one of them. With more than half of these structures being reserved for food, energy, materials processing, water storage, sewage treatment, manufacturing, and other purposes, we can accommodate an eventual self-sustaining population of about two and a half million. When we started our journey to the Epsilon Eridani solar system, many of these living spaces were left empty to accommodate population growth. In fact, when my wife and son and I left Earth one hundred ten years ago, we began with a population of only two hundred and fifty thousand. During the trip we've grown to more than six hundred thousand. So, we have more than enough space to accommodate a growing population for our one hundred ten–year trip back to planet Earth. I'm sorry to say that our medical advancements haven't progressed to where I'll live to see that day.

If you think our spacecraft is impressive, wait till you learn about how the living space structures are designed. It will blow your mind. First of all, the interior space has an ultra-high-definition hologram at the far end that can project anything from a Pacific Ocean scene to a view of a New Hampshire mountain range. In the foreground trees and plants that are indigenous to the holograph scene are arranged to enhance the perceived reality of the view. Further enhancement for a Pacific Ocean view can be provided by adding salt air, the sound of crashing waves, and seagulls. Best of all, the holograph and foreground plantings can be changed occasionally for variety.

Notice the uniquely styled house located at the back end of the living space. Generally speaking, most people prefer an upside-down house where the bedrooms are on the bottom floor and the living area is on top. This of course maximizes the magnificent views. With a one-hundred-foot-high ceiling, LED lighting, and holographic imaging the simulated sky looks very real. Moreover, variations in weather can be programmed as desired.

Now, here is the best part. Each family is given an android. The android is fueled by nuclear fusion and can perform almost any function that humans are able to do. And —you guessed it —they do everything from cooking the meals to manicuring the garden areas. One interesting and important aspect regarding the androids is their ability to repair and replicate themselves.

I think you can understand why Earth's population has been migrating to these living spaces in droves. With an abundance of helium 3 fuel, construction materials, and programmed robotic labor, the living spaces and material accommodations are essentially free. It's a form of utopia, but the humans need to supervise android activities and contribute to society. Obviously, they have time for raising children, improving their education, and other things that in the past was limited by working for a living.

NOTE: I could have stopped at this point in the Epsilon Eridani story, but I wanted to include excerpts from one other chapter because I think you'll find it to be quite interesting and entertaining. It illustrates the contrast between primitive and advanced technologies, and how people's lives and thinking are influenced. This part of the story begins at a point in time when the Epsilon 1 spacecraft is ready to return to Earth, and a family from the Epsilon planet's inhabitants has agreed to return with them. The Epsilon family is partially comprised of Jason Harvath's son, Stewart, who during an encounter with the Epsilon people married a girl named Wanaka and had a baby. Wanaka is the granddaughter of the planet's leader, Homanz. Included in the Epsilon family were Wanaka's immediate family and relatives, including Homanz. The people who lived on the Epsilon planet were quite primitive—similar to ancient Egypt—and at this time in the story, had not been exposed to Epsilon 1's technology:

On their arrival back to Epsilon 1, they were greeted by a large crowd of cheering people and marching bands. Homanz, Wanaka, and others in the family were surprised and pleased by this welcoming. The warm greeting allowed the Epsilon 1 population to publicly demonstrate their acceptance. In fact, the Epsilon 1 population was anxious to see the family's reaction when they saw where they were going to live and the technologies that they were about to experience.

Although Jason did not approve of having the press intrude on the privacy of the family, he realized they had become celebrities. And, unless the public saw them at this time, they wouldn't stop hounding him until they did. So, he reluctantly agreed.

The plan was for the family to visit Jason's home module as a first step.

In preparation for this, the press was allowed to set up cameras at hidden locations. The fact is, Jason, his wife, Jennifer, and Stewart were also anxious to see the family's reaction. Sort of like seeing the reaction of a child opening a Christmas gift.

It was planned that the arrival at Jason's home module would take place at mid-afternoon, and after describing what the family would see in the home module, they would spend the night. The next step was for Stewart to show Wanaka their home module the next day, with other family members going to their home modules at the same time. Jason and Stewart had done a lot of work helping to design the new modules to include replicas of familiar furniture and other items so the families would feel at home. The landscaping was made to look like their former home planet, and the holographic imaging was recorded from surroundings that were familiar to them.

Because a fence had been set up to block the Epsilon people from seeing the Epsilon 1's common area while they were being rescued from their planet's flood, this was the first time for the family to see the wonders of a man-made paradise. Although the people on board Epsilon 1 had become used to the common area, to a newcomer, the views would be breathtaking.

Using his translation device, Homanz asks Jason, "Why you have kept this from our people?"

"Homanz, we made a decision at the beginning of the rescue to not expose your people to our technology. There is an old saying that what you don't know can't hurt you or make you envious. If the Epsilon people knew of our technology, they would probably want it for themselves, and it would make their lives less content."

"I understand, Jason, but this tells me you not know true nature my people."

"Homanz, you are probably right, but first we need to travel to my home on board this train."

While on the maglev train, the Epsilon family expressed amazement at the quiet speed and luxury they were experiencing. Upon arrival at Jason's home coordinates, they traveled by way of an elevator to Jason's home and entered what appeared to be a wonderland. At one end of the 800-foot diameter by 100-foot-high living space was a magnificent house with beautiful landscaping. On each side of the house stood detached buildings. At the far end, opposite the house, a magnificent ultra-high-definition holographic view of the Pacific Ocean could be seen. Aside from the actual size of the living space, everything appeared to be infinitely expansive. The ocean view overlooked a high cliff with tables, chairs, fire pit, barbeque, and kitchen appliances arranged for dining and relaxing activities. From this cliff side vantage point, a person could smell salt air and listen to the sounds of crashing waves and seagulls. Between the house and cliff were perfectly manicured pathways, gardens, and trees, intrinsic to Earth's Pacific coast. At the center of the home module was a large, unique swimming pool. It had a white sand bottom and two bridges passing over streams that connected two smaller pools.

As the Epsilon people looked in awe at the magnificent sight of Jason and Jennifer's home module, an android drove up in an open electric cart. The android asked six of the group to get in the cart for the trip to the house. He said that he would come back for the remaining four on a second trip. Stewart, Wanaka, and Wanaka's mother and father stayed behind and advised the android that they would walk to the house.

The house was well appointed and functional. It was an upside-down house in that five bedrooms, each with its own bath, were located on the lower level, while the kitchen, dining room, and living area were located on the top floor to take advantage of the views.

In preparation for the visit, the android named Joe had prepared the five bedrooms to accommodate the family members, however, because the bedroom space was limited, the plan was for Jason and Jennifer to spend the night in the poolside cabana and recreation room.

After showing each family member their accommodations, Joe told them to meet by the oceanside cliff for dinner. He had prepared a dinner that included recipes made with ingredients from their planet to help make the guests feel more at home. Before dinner, everyone was offered wine and a chance for conversation. When they had all gathered, Jason stepped forward to speak to the group.

"I know each of you has questions, so this is a good time to ask."

Wanaka stood and asked, "Will each our families have place like this?"

"Yes. Stewart, Jennifer, and I have spent much time designing each of your home modules, and I hope you like them. Homanz will live in his son and daughter-in-law's home module but will have a separate house. Our plan is for each of you to visit your home module tomorrow."

Wanaka's mother said, "I see many things but not know about them. Please explain how work."

"Yes, let me begin. They work using what we call electric power. Electric power, for all that you see, is provided by special boxes that receive fuel from pipes. This fuel is like coal or firewood, but we call it helium 3. In addition to the helium 3 fuel, we also receive air and water from pipes. Waste and trash are processed in each home module and sent in special containers to a central location using a lifting device like the one you came here in. We call it a service elevator. The android Joe removes trash each day. Special containers of food and other necessities come into the module using this same service elevator. Sensors notify a device to automatically order items we are running out of. For instance, if we need a container of salt, the salt holder senses that it needs to be replaced. It signals to the ordering device to place the order. When the new container arrives, Joe places it into one of the salt container holders and removes the empty container and sends it back for reprocessing. The empty salt container is then cleaned and returned to the salt refill location. In other words, all food containers are reused so there is almost no waste."

With a puzzled look on his face, Wanaka's father asks, "how you move house and big things to what you call home module?"

"We use a third elevator for big things. Our house was made of smaller parts and fitted together in here. We design all large things to fit into the large elevator. These parts are also designed for disassembly after the house becomes old. We use this method for almost all the things you see, so everything is reused. When disassembled house parts cannot be reused to remake a new house. The materials are reworked for other purposes."

Wanaka's cousin pointed to the android. "How Joe work?"

"Joe is like a human; except he is mechanical and operates using electricity. Joe can work twenty-four hours per day. During the day he does what you see him doing now. But at night he spends most of his time working on our landscape. When flowers die, he replaces them. When leaves on trees die, he clips them off. When weeds grow, he plucks them out. When grass is too long, he cuts it. When trees get too tall, he uses a machine to cut the branches. When trees die, he orders new trees and cuts the old tree for making wood products and mulch."

Wanaka stepped forward and asks, "How look like ocean?"

Jason smiled as they gazed at the scene. "It is an artificial image, but as you can see it looks very real. Jennifer and I like this image very much, but we are planning to have a new one soon."

"What new image like?"

"Jennifer and I have decided to do something like one of our neighbors. It's an image that simulates the four seasons on our home planet that we call spring, summer, fall, and winter. It will be a mountain view from a place called New Hampshire. Since we have almost unlimited fuel for electric power, our home modules can simulate temperatures and conditions without concern for running out. So, in the four-season image, we will have lower temperatures in the spring, winter, and fall, including rain in the spring and snow in the winter. Because our current landscape design is for a mild climate, we will replace the palm trees with maple trees and grassy meadows. Even though we can easily create the artificial image now, we need to spend time modifying the landscape to match the image."

Homanz wrinkled his brow. "What people do for work?"

Jason nodded. "Good question, everyone who can work is required to work two days each week but can work more if they want to. Almost all manual labor is done by androids like Joe. So almost all human jobs are to supervise androids. Everyone receives the same pay credits whether they work more hours or have a highly skilled profession like a doctor. Since everyone can have almost anything that they want, a pay credit limit prevents abuse."

Homanz asks, "How you decide good worker from poor worker?"

"Homanz, that's a very good question. We have twelve sectors. Each one has a large city, or what you might call a village. Each sector has a yearly competition for awards. Awards can be for excellent products or service and includes movie production, sports, and school competition. The top sector gets a special recognition each year. This recognition includes having each family getting extra credits toward things like a more frequent remodeling of their home module."

Wanaka sets down her drink. "So, what people do when not working?" she asks.

"Wanaka, we work to better ourselves. What I mean is we learn as much as we can and do creative things. We also exercise and play sports to improve our health."

"What you do?"

"As you can see, I have what I call a creativity workshop located to the right side of our house. In this workshop, I have woodworking and metalworking tools. I also have an exercise room and a place to write books. To the left side of the house is another creativity workshop for Jennifer. Jennifer likes to grow things like flowers and vegetables. She also likes to compose music and paint."

"What Stewart do?" Wanaka asks.

"I can't speak for Stewart, but I know that he plans to spend one more year at the university to get his doctorate in science. After that he will probably work at the central science laboratory with me. And if I know Stewart, he will become consumed in his work, and like me he will work more than just two days a week. What do you say, Stewart?"

Stewart pauses for a moment. "You're right about going back to the university, but I don't know about being consumed by my work. Right now, I want to spend as much time as I can with my family."

Jason looks around the room. "Are there any more questions before we have our meal?"

Wanaka's cousin raises his hand. "What kind jobs we do?" he asks.

"There are many types of jobs from which to choose. Since you are familiar with hunting, farming, and fishing, you might choose to supervise one of our wild animal preserves, cat-

tle ranches, food growing facilities, or
fisheries. Ok let's sit down to eat. We
can talk more then."

In this story I describe how humans might sustain a
growing population by living in space and eventually traveling beyond our solar system in multi-generational spacecraft. From a biblical perspective much of
what I've proposed might happen during and/or after
the Millennium.

In this regard, I thought of how this story could be
made into a movie that would illustrate how what I'm
saying might be possible. Like ants and bees that relentlessly pursue a single task, I would show how robots
could make what I'm saying happen. Isn't this what
we could do with millions—or billions—of self-repli-
cating and programmable robots working in space and
building huge satellite habitats/spacecraft? In the movie
I would show a zero-gravity space factory in which
materials from asteroids and moons could be sorted into
their various types. Using these materials, the self-rep-
licating robots would be shown making, among other things, tunnel boring machines and cargo carrying
spacecraft. Included in my proposed movie I would
also illustrate how millions of cargo carrying spacecraft
and boring machines would be relentlessly excavating

asteroids and moons and returning materials back to space factories. This image would resemble ants, or bees, going to and from their assigned objectives.

The point I'm trying to make is that overpopulation of planet Earth need not be a problem. With the potential materials, water, and helium 3 fuel available in just our solar system, housing 30 trillion people in luxurious living conditions might be possible.

Finally, with almost all of Earth's inhabitants living in space habitats and having the ability to communicate and debate big issues—like the origin of life, religious beliefs, and human purpose—a commonly held truth might emerge. I have no doubt that this emerged truth would result in almost all people, alive during that time, devoting their human lives to Jesus Christ and expecting eternal life in God's Kingdom of Heaven. However, this is not what we see happening today. With political correctness, media bias, corrupted entertainment, atheism, secular immorality, multiple religious beliefs, cultural differences, language problems, and a majority of Earth's people struggling just to survive, we currently have chaos. As a result, communication, debate, and logic generally falls on deaf ears, or in the case of third World people, on no ears at all.

So, there you have it. I've reported. Now you decide.

About the Author

M r. Bongaards' degree is in mechanical engineering from Northeastern University. Upon graduation, in 1964, he was commissioned a lieutenant in the United States Army and served in Vietnam.

In August 1966 he began his career with the Westinghouse Electric Corporation. During that time, he worked on nuclear steam generator development and was eventually promoted to section engineering manager and later, department engineering manager.

In 1982, he transferred to the Thermo King Corporation (a subsidiary of Westinghouse) to become engineering manager for the Truck Transport Refrigeration Equipment Department. During that time, he was given responsibility for design engineering activities at Thermo King factories in Barcelona, Spain; Hamble, England; and Prague, Czech Republic.

Mr. Bongaards has written numerous technical papers for international conferences and for the American Society of Mechanical Engineers (ASME) and currently holds seven patents. He became chairman of the Florida West Coast Section of ASME in 197475 and passed the Florida professional engineering examination in 1975. This is the fifth book that he has written since his retirement in 2001.